THIRTY YEARS
THAT SHOOK PHYSICS

The Story of Quantum Theory

GEORGE GAMOW

Illustrations by the Author

DOVER PUBLICATIONS, INC.
NEW YORK

Published in Canada by General Publishing Company, Ltd., 30 Lesmill Road, Don Mills, Toronto, Ontario.
Published in the United Kingdom by Constable and Company, Ltd., 10 Orange Street, London WC2H 7EG.

This Dover edition, first published in 1985, is an unabridged and unaltered republication of the work first published by Doubleday & Co. Inc., New York, in 1966.

Manufactured in the United States of America
Dover Publications, Inc., 31 East 2nd Street, Mineola, N.Y. 11501

Library of Congress Cataloging in Publication Data

Gamow, George, 1904–1968.
Thirty years that shook physics.

Reprint. Originally published: Garden City, N.Y. : Doubleday, 1966.
Includes bibliographical references and index.
1. Quantum theory.   2. Physics—History.   I. Title.
QC174.12.G35   1985   530.1'2'09   85-6797
ISBN 0-486-24895-X

TO THE FRIENDS OF MY YOUTH

L. Landau   A. Bohr   G. Gamow   E. Bohr   E. Teller

(Photographed probably by Dr. H. A. Casimir)   Copenhagen 1931

LIST OF PLATES

BIOGRAPHICAL PREFACE

Thirty Years That Shook Physics displays Dr. Gamow's artistic gift as well as his ability to expound science in the layman's language. Dr. Gamow himself has acknowledged Sandro Botticelli as his master in portraiture, and for any art students who may happen upon this book it will make an interesting subsidiary exercise to find the Botticelli influence in the studies of Max Planck (page 6) and Niels Bohr (page 29). The further relation of Dr. Gamow's style to the Pop Art of recent notoriety will be more readily apparent. The philosophically minded may find it significant that continuity in the stream of art from the Italian Renaissance to the mid-twentieth century's Madison Avenue should be expressed in the works of a mathematical physicist celebrated for his development of the Big Bang theory of cosmic creation.

Even if they could write with comparable flavor and had equal mastery of the difficult science, few (if any) of today's physicists could have produced a book like this one. It is a retrospective view of a crucial period of intellectual development, written by one who was there and who took active part. The great names you will find in the following pages were more than names to Dr. Gamow; these scientific giants, who remade the universe in man's mind, were his teachers, his friends, and his colleagues. Because of the cosmopolitan circumstances of his life, it was his great good luck to have been among those in attendance at many of the momentous events of the thirty years that shook physics.

Dr. Gamow was born on March 4, 1904, in Odessa, Russia. In early youth he turned to science and spent a year studying paleontology. This experience, he said later, equipped him "to tell a dinosaur from a cat by the shape of the little toes." He entered the University of Leningrad, from which he received a Ph.D. degree in 1928, and spent a year at the University of Göttingen, in Germany, on a traveling fellowship. In 1928–29 he worked with Niels Bohr in Copenhagen and in 1929–30 with Ernest Rutherford at the Cavendish Laboratory, Cambridge, England.

Dr. Gamow was twenty-four when he made his first major contribution to physical theory. Concurrently, but independently, he, on the one hand, and the American physicist E. U. Condon and the British physicist R. W. Gurney, on the other, explained the emission of alpha-particles from radioactive atoms by applying to the process the then new methods of wave mechanics. Two years later, in 1930, he made the successful prediction that protons would be more useful than alpha-particles in the experiments popularly known as "atom-smashing," and in the same year he suggested the liquid drop model for the nuclei of heavy elements. In 1929 he collaborated with R. Atkinson and F. Houtermans in formulating the theory that the sun's heat and light resulted from thermonuclear processes, and his theory of the origin of chemical elements through neutron capture dominated cosmological thinking at one period in the 1940s. He also has contributed to the fundamentals of biology, having proposed that the four nucleotides of the DNA molecule compose a code whose different combinations act as templates in the organization of the various amino acid molecules.

Dr. Gamow's personal characteristics are almost as

formidable as his creative achievements. A giant, six feet three and well over 225 pounds, he is given to puckish humor, as readers of his *Mr. Tompkins* fantasies well know. When he and his student, R. Alpher, signed their names to the preliminary calculations of their paper, *The Origin of Chemical Elements,* in 1948, Gamow commented, "Something is missing," and, crediting Hans Bethe *in absentia,* made the signature "Alpher, Bethe and Gamow." He speaks six languages and is a frequent and popular lecturer with a heavily accented delivery that moved a friend to observe that the six languages were all different dialects of one language—"Gamovian." Traces of Gamovian creep into his literary style from time to time, but the editor who would expunge them ruthlessly would be a pedant of the worst kind, insensitive to individual enrichment of the language.

Dr. Gamow's ability as a linguist, however accented, reflects the ground he has covered in his professional career. After his studies with Bohr and Rutherford, he returned to Russia as Master in Research at the Academy of Sciences in Leningrad but left his native land for good in 1933. He lectured in Paris and London and at the University of Michigan summer school, then joined the faculty of George Washington University, Washington, D.C., where he was professor of physics from 1934 to 1956. He became a United States citizen in 1940 and acted as a Navy, Army, Air Force and Atomic Energy Commission consultant during and after World War II. Since 1956 he has been on the faculty of the University of Colorado, Boulder.

Dr. Gamow has written many technical papers and one technical book, *Atomic Nucleus* (Oxford University Press, 1931, revised 1937 and 1949). His popular writing includes numerous *Scientific American* articles and the following books:

Mr. Tompkins in Wonderland, Cambridge University Press, 1939

Mr. Tompkins Explores the Atom, Cambridge University Press, 1943

Mr. Tompkins Learns the Facts of Life, Cambridge University Press, 1953

Atomic Energy in Cosmic and Human Life, Cambridge University Press, 1945

The Birth and Death of the Sun, Viking Press, 1941

Biography of the Earth, Viking Press, 1943

One, Two, Three . . . Infinity, Viking Press, 1947

Creation of the Universe, Viking Press, 1952

Puzzle-Math (with M. Stern), Viking Press, 1958

The Moon, H. Schuman, 1953

Matter, Earth and Sky, Prentice-Hall, 1958 (2nd Edition, 1965)

Physics: Foundation and Frontiers (with J. Cleveland), Prentice-Hall, 1960

Atom and Its Nucleus, Prentice-Hall, 1961

Biography of Physics, Harper and Brothers, 1961

A Star Called the Sun, Viking Press, 1965

A Planet Called the Earth, Viking Press, 1965

He took up illustrating for the second *Mr. Tompkins* book when World War II interrupted communication between him and the English artist who had worked with him on the earlier book of the series. In 1956 he received the Kalinga Prize from UNESCO for his popular interpretations of science for lay readers.

Dr. Gamow was a member of the Academy of Science of the U.S.S.R. until, as he says, he was "fired after leaving Russia." He is a member of the Royal Danish Academy of Sciences and the National Academy of Sciences of the United States.

John H. Durston

PREFACE

Two great revolutionary theories changed the face of physics in the early decades of the twentieth century: the *Theory of Relativity* and the *Quantum Theory*. The former was essentially the creation of one man, Albert Einstein, and came in two installments: the Special Theory of Relativity, published in 1905, and the General Theory of Relativity, published in 1915. Einstein's Theory of Relativity called for radical changes in the classical Newtonian concept of space and time as two independent entities in the description of the physical world, and led to a unified four-dimensional world in which time is regarded as the fourth coordinate, though not quite equivalent to the three space coordinates. The Theory of Relativity introduced important changes in the treatment of the motion of electrons in an atom, the motion of planets in the solar system, and the motion of stellar galaxies in the universe.

The Quantum Theory, on the other hand, is the result of the creative work of several great scientists starting with Max Planck, who was the first to introduce into physics the notion of a quantum of energy. The theory went through many evolutionary stages and gives us today a deep insight into the structure of atoms and atomic nuclei as well as that of bodies of the sizes familiar to our everyday experience. As of today Quantum Theory is not yet completed, especially in its relation to the Theory of Relativity and the problem of elementary particles, being stalled (temporarily) by tremendous difficulties encountered on the way toward further development.

It is the development of the Quantum Theory that this book will discuss. The author was first introduced to the idea of quanta and Bohr's atomic model at the age of eighteen when he enrolled as a student in the University of Leningrad, and later, at the age of twenty-four, he had the good luck to become Bohr's student in Copenhagen. During those memorable years at *paa Blegdamsvej* (the address of Bohr's Institute) he had the opportunity of meeting many scientists who contributed to the early development of the Quantum Theory, and of taking part in their discussions. The account that follows is an outgrowth of those experiences, centered on the great and lovable figure of Niels Bohr. The author hopes that the new generation of physicists will find some interesting information in the pages that follow.

G. Gamow

January 1965   George Gamow

CONTENTS

INTRODUCTION

The opening of the twentieth century heralded an unprecedented era of turnover and re-evaluation of the classical theory that had governed Physics since pre-Newtonian times. Speaking on December 14, 1900, at the meeting of the German Physical Society, *Max Planck* stated that paradoxes pestering the classical theory of the emission and absorption of light by material bodies could be removed if one assumed that *radiant energy can exist only in the form of discrete packages.* Planck called these packages *light quanta.* Five years later, *Albert Einstein* successfully applied the idea of light quanta to explain the empirical laws of photoelectric effect; that is, the emission of electrons from metallic surfaces irradiated by violet and ultraviolet light. Still later, *Arthur Compton* performed his classical experiment, which showed that the scattering of X-rays by free electrons followed the same law as the collision between two elastic spheres. Thus, within a few years the novel idea of quantization of radiant energy firmly established itself in both theoretical and experimental physics.

In the year 1913, a Danish physicist, *Niels Bohr,* extended Planck's idea of quantization of radiant energy to the description of mechanical energy of electrons within an atom. Introducing specific "quantization rules" for the mechanical systems of atomic sizes, he achieved a logical interpretation of Ernest Rutherford's planetary model of an atom, which rested on a solid experimental basis but on the other side stood in sharp contradiction to all the fundamental concepts of classical physics. Bohr calculated the energies of various discrete quantum states of atomic electrons and interpreted the emission of light as the ejection of a light quantum with energy equal to the energy difference between the initial and final quantum states of an atomic electron. With his calculations he was able to explain in great detail the spectral lines of hydrogen and heavier elements, a problem which for decades had mystified the spectroscopists. Bohr's first paper on the quantum theory of the atom led to cataclysmic developments. Within a decade, due to the joint efforts of theoretical as well as experimental physicists of many lands, the optical, magnetic, and chemical properties of various atoms were understood in great detail. But as the years ran by, it became clearer and clearer that, successful as Bohr's theory was, it was still not a final theory since it could not explain some things that were known about atoms. For example, it failed completely to describe the transition process of an electron from one quantum state to another, and there was no way of calculating the intensities of various lines in optical spectra.

In 1925, a French physicist, *Louis de Broglie,* published a paper in which he gave a quite unexpected interpretation of Bohr quantum orbits. According to de Broglie, the motion of each electron is governed by some mysterious *pilot waves,* whose propagation velocity and length depend on the velocity of the electron in question. Assuming that the length of these pilot waves is inversely proportional to the electron's velocity, de Broglie could show that various quantum orbits in Bohr's model of the hydrogen atom were those that could accommodate an *integral number* of pilot waves. Thus, the model of an atom began to look like some kind of musical instrument with a basic tone (the innermost orbit with the lowest energy) and various overtones (outlying orbits with higher energy). One year after their publication, de Broglie's ideas were extended and brought into more exact mathematical form by the Austrian physicist *Erwin Schrödinger*, whose theory became known as *Wave Mechanics*. While explaining all the atomic phenomena for which Bohr's theory already worked, wave mechanics also explained those phenomena for which Bohr's theory failed (such as the intensities of spectral lines, etc.), and in addition predicted some new phenomena (such as diffraction of an electron beam) which had not even been dreamed of, either in classical physics or in Planck-Bohr quantum theory. In fact, wave mechanics provided a complete and perfectly self-consistent theory of all atomic phenomena, and, as was shown in the late twenties, could explain also the phenomena of radioactive decay and artificial nuclear transformations.

Simultaneously with Schrödinger's paper on wave mechanics, there appeared a paper of a young German physicist, *W. Heisenberg,* who developed the treatment of quantum problems by using the so-called "non-commutative algebra," a mathematical discipline in which $a \times b$ is not necessarily equal to $b \times a$. The simultaneous appearance of Schrödinger's and Heisenberg's papers in two different German magazines (*Ann. der Phys.* and *Zeitsch. der Phys.*) astonished the world of theoretical physics. These two papers looked as different as they could be, but led to exactly the same results concerning atomic structure and spectra. And it took more than a year until it was found that the two theories were physically identical except for being expressed in two entirely different mathematical forms. It was as if America was discovered by Columbus, sailing westward across the Atlantic Ocean, and by some equally daring Japanese, sailing eastward across the Pacific Ocean.

But there still remained one sharp thorn in the crown of the Quantum Theory, and it made itself felt painfully whenever one tried to quantize mechanical systems which, because of the very high velocities involved (close to the speed of light) required relativistic treatment. Many unsuccessful attempts had been made to unite the Theory of Relativity with the Theory of Quanta until finally, in 1929, a British physicist, *P. A. M. Dirac,* wrote his famous *Relativistic Wave Equation.* The solutions of this equation gave a perfect description of the motion of atomic electrons at velocities close to that of light, and gave automatically, as an unexpected bonus, the explanation of their linear and angular mechanical momenta and magnetic moments. Some formal difficulties connected with handling this equation led Dirac to suggest that along with ordinary negatively charged electrons *there must also exist positively charged anti-electrons.* His prediction was brilliantly verified a few years later when anti-electrons were found in the cosmic rays. The theory of *anti-particles* was extended to elementary particles other than electrons, and today we have anti-protons, anti-neutrons, anti-mesons, etc.

Thus, by 1930, only three decades after Planck's momentous announcement, the Quantum Theory took the final shape with which we are now familiar. Very little theoretical progress was made in the decades that followed these breathtaking developments. On the other hand, these later years have been quite fruitful in the field of experimental studies, especially in the investigation of the numerous newly discovered elementary particles. We are still waiting for a breakthrough in the solid wall of difficulties which prevent us from un-

Sir J. J. Thomson and Lord Rutherford.
(Photographer unknown)

derstanding the very existence of elementary particles, their masses, charges, magnetic moments, and inter-actions. There is hardly any doubt that when such a breakthrough is achieved, it will involve concepts that will be as different from those of today as today's con-cepts are different from those of classical physics.

In the following chapters an attempt will be made to describe the growth of the Quantum Theory of energy and matter through the first thirty years of its turbulent development, stressing the conceptual differences be-tween "good old" classical physics and the new look physics has assumed in the twentieth century.

CHAPTER I

M. PLANCK AND LIGHT QUANTA

The roots of Max Planck's revolutionary statement that light can be emitted and absorbed only in the form of certain discrete energy packages goes back to much earlier studies of Ludwig Boltzmann, James Clerk Maxwell, Josiah Willard Gibbs, and others on the statistical description of the thermal properties of material bodies. The Kinetic Theory of Heat considered heat to be the result of random motion of the numerous individual molecules of which all material bodies are formed. Since it would be impossible (and also purposeless) to follow the motion of each single individual molecule participating in thermal motion, the mathematical description of heat phenomena must necessarily use statistical method. Just as the government economist does not bother to know exactly how many acres are seeded by farmer John Doe or how many pigs he has, a physicist does not care about the position or velocity of a particular molecule of a gas which is formed by a very large number of individual molecules. All that counts here, and what is important for the economy of a country or the observed macroscopic behavior of a gas, are the averages taken over a large number of farmers or molecules.

One of the basic laws of *Statistical Mechanics,* which is the study of the average values of physical properties for very large assemblies of individual particles involved in random motion, is the so-called *Equipartition Theorem,* which can be derived mathematically from the Newtonian laws of Mechanics. It states that: *The total energy contained in the assembly of a large number of individual particles exchanging energy among themselves through mutual collisions is shared equally (on the average) by all the particles.* If all particles are identical, as for example in a pure gas such as oxygen or neon, all particles will have on the average equal velocities and equal kinetic energies. Writing E for the total energy available in the system, and N for the total number of particles, we can say that the average energy per particle is E/N. If we have a collection of several kinds of particles, as in a mixture of two or more different gases, the more massive molecules will have the lesser velocities, so that their kinetic energies (proportional to the mass and the square of the velocity) will be on the average the same as those of the lighter molecules.

Consider, for example, a mixture of hydrogen and oxygen. Oxygen molecules, which are 16 times more

massive than those of hydrogen, will have average velocity $\sqrt{16} = 4$ times smaller than the latter.†

While the equipartition law governs the *average distribution of energy* among the members of a large

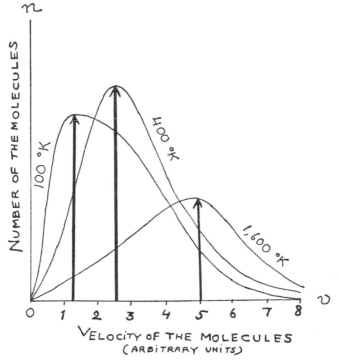

Fig. 1. Maxwell's distribution: the number of molecules
having different velocities v is plotted against the velocities
for three different temperatures, 100°, 400°, and 1600°K.
Since the number of molecules in the container remains
constant, the areas under the three curves are the same.
The average velocities of the molecules increase proportionally to the square root of the absolute temperature.

† Since kinetic energy is the product of [mass] × [velocity]²,
this product will remain the same if the mass increases by a
factor 16 and velocity decreases by a factor 4. In fact, $4^2 = 16$!

assembly of particles, the velocities and energies of individual particles may deviate from the averages, a phenomenon known as *statistical fluctuations*. The fluctuations can also be treated mathematically, resulting in curves showing the relative number of particles having velocities greater or less than the average for any given temperature. These curves, first calculated by J. Clerk Maxwell and carrying his name, are shown in Fig. 1 for three different temperatures of the gas. The use of the statistical method in the study of thermal motion of molecules was very successful in explaining the thermal properties of material bodies, especially in the case of gases; in application to gases the theory is much simplified by the fact that gaseous molecules fly freely through space instead of being packed closely together as in liquids and solids.

STATISTICAL MECHANICS AND THERMAL RADIATION

Toward the end of the nineteenth century Lord Rayleigh and Sir James Jeans attempted to extend the statistical method, so helpful in understanding thermal properties of material bodies, to the problems of thermal radiation. All heated material bodies emit electromagnetic waves of different wavelengths. When the temperature is comparatively low—the boiling point of water, for example—the predominant wavelength of the emitted radiation is rather large. These waves do not affect the retina of our eye (that is, they are invisible) but are absorbed by our skin, giving the sensation of warmth, and one speaks therefore of heat or *infrared radiation*. When the temperature rises to about 600°C (characteristic of the heating units of an electric range) a faint red light is seen. At 2000°C (as in the filament of an electric bulb) a bright white light which contains all the wavelengths of the *visible radia-*

tion spectrum from red to violet is emitted. At the still higher temperature of an electric arc, 4000°C, a considerable amount of invisible *ultraviolet radiation* is emitted, the intensity of which rapidly increases as the

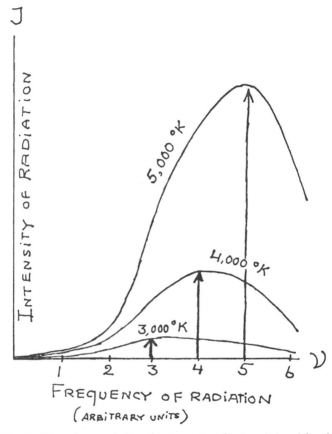

Fig. 2. *The observed distribution of radiation intensities for different frequencies* ν *is plotted against the frequencies. Since the radiation energy content per unit volume increases as the fourth power of the absolute temperature* T, *the areas under the curves increase. The frequency corresponding to maximum intensity increases proportionally to the absolute temperature.*

temperature rises still higher. At each given temperature there is one predominant vibration frequency for which the intensity is the highest, and as the temperature rises this predominant frequency becomes higher and higher. The situation is represented graphically in Fig. 2, which gives the distribution of intensity in the spectra corresponding to three different temperatures.

Comparing the curves in Figs. 1 and 2, we notice a remarkable qualitative similarity. While in the first case the increase of temperature moves the maximum of the curve to higher molecular velocities, in the second case the maximum moves to higher radiation frequencies. This similarity prompted Rayleigh and Jeans to apply to thermal radiation the same Equipartition Principle that had turned out to be so successful in the case of gas; that is, to assume that the total available energy of radiation is distributed equally among all possible vibration frequencies. This attempt led, however, to catastrophic results! The trouble was that, in spite of all similarities between a gas formed by individual molecules and thermal radiation formed by electromagnetic vibrations, there exists one drastic difference: while the number of gas molecules in a given enclosure is always finite even though usually very large, the number of possible electromagnetic vibrations in the same enclosure is always infinite. To understand this statement, one must remember that the wave-motion pattern in a cubical enclosure, let us say, is formed by the superposition of various standing waves having their nodes on the walls of the enclosure.

The situation can be visualized more easily in a simpler case of one-dimensional wave motion, as of a string fastened at its two ends. Since the ends of the string cannot move, the only possible vibrations are those shown in Fig. 3 and correspond in musical terminology to the fundamental tone and various overtones

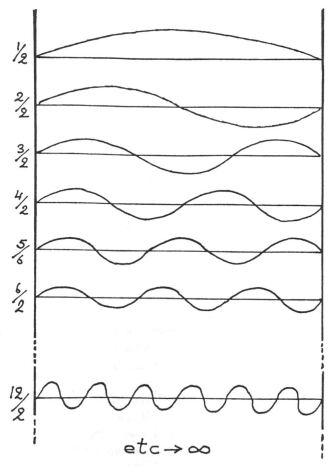

½

²⁄₂

³⁄₂

⁴⁄₂

⁵⁄₆

⁶⁄₂

¹²⁄₂

etc → ∞

Fig. 3. The basic tone and higher overtones in the case of the one-dimensional continuum—for example, a violin string.

of the vibrating string. There may be one half-wave on the entire length of the string, two half-waves, three half-waves, ten half-waves, . . . a hundred, a thousand, a million, a billion . . . any number of half-waves. The corresponding vibration frequencies of various overtones will be double, triple . . . tenfold, a hundredfold, a millionfold, a billionfold . . . etc., of the basic tone.

In the case of standing waves within a three-dimensional container, such as a cube, the situation will be similar though somewhat more complicated, leading to unlimited numbers of different vibrations with shorter and shorter wavelengths and correspondingly higher and higher frequencies. Thus, if E is the total amount of radiant energy available in the container, the Equipartition Principle will lead to the conclusion that each individual vibration will be allotted E/∞, an infinitely small amount of energy! The paradoxicalness of this conclusion is evident, but we can point it even more sharply by the following discussion.

Suppose we have a cubical container, known as "Jeans' cube," the inner walls of which are made of ideal mirrors reflecting 100 per cent of the light falling on them. Of course, such mirrors do not exist and cannot be manufactured; even the best mirror absorbs a small fraction of the incident light. But we can use the notion of such ideal mirrors in theoretical discussions as the limiting case of very good mirrors. Such reasoning, whereby one *thinks* what would be the result of an experiment in which ideal mirrors, frictionless surfaces, weightless bars, etc., are employed, is known as a "thought experiment" (*Gedankenexperiment* is the original term), and is often used in various branches of theoretical physics. If we make in a wall ·of Jeans' cube a small window and shine in some light, closing the ideal shutter after that operation, the light will stay in for an indefinite time, being reflected to and fro from the ideal mirror walls. When we open the shutter sometime later we will observe a flash of the escaping light. The situation here is identical in principle to pumping some gas into a closed container and letting it out again later. Hydrogen gas in a glass container can stay indefinitely, representing an ideal case. But hydrogen will not stay long in a container made of palladium metal, since hydrogen molecules are known to diffuse rather easily through this material. Nor can one use a glass container for keeping hydrofluoric acid, which reacts chemically with glass walls. Thus, Jeans' cube with the ideal mirror walls is after all not such a fantastic thing!

There is, however, a difference between the gas and the radiation enclosed in a container. Since the molecules are not mathematical points but have certain finite diameters, they undergo numerous mutual collisions in which their energy can be exchanged. Thus, if we inject into a container some hot gas and some cool gas, mutual collisions between the molecules will rapidly slow down the fast ones and speed up the slow ones, resulting in even distribution of energy in accordance to the Equipartition Principle. In the case of an ideal gas formed by point-molecules, which of course does not exist in nature, mutual collisions would be absent and the hot fraction of the gas would remain hot while the cool fraction would remain cool. The exchange of energy between the molecules of an ideal gas can be stimulated, however, by introducing into the container one or several particles with finite though small diameters (Brownian particles). Colliding with them, fast point-sized molecules will communicate to them their energy, which will be communicated in turn to the other slower point-sized molecules.

In the case of light waves the situation is different, since two light beams crossing each other's path do not affect each other's propagation in any way.‡ Thus,

‡ To avoid an objection on the part of those readers who know much more than necessary for understanding this discussion, the author hastens to state that, according to modern quantum electrodynamics, some scattering of light by light must be ex-

to procure the exchange of energy between the standing waves of different lengths, we must introduce into the container small bodies that can absorb and re-emit all possible wavelengths, thus permitting energy exchange among all possible vibrations. Ordinary black bodies, such as charcoal, have this property, at least in the visible part of the spectrum, and we may imagine "ideal black bodies" which behave in the same way for all possible wavelengths. Placing into Jeans' cube a few particles of ideal coal dust, we will solve our energy-exchange problem.

Now let us perform a thought experiment, injecting into an originally empty Jeans' cube a certain amount of radiation of a given wavelength—let us say some red light. Immediately after injection, the interior of the cube will contain only red standing waves extending from wall to wall, while all other modes of vibrations will be absent. It is as if one strikes on a grand piano one single key. If, as it is in practice, there is only very weak energy exchange among different strings of the instrument, the tone will continue to sound until all the energy communicated to the string will be dissipated by damping. If, however, there is a leak of energy among the strings through the armature to which they are attached, other strings will begin to vibrate too until, according to the Equipartition Theorem, all 88 strings will have energy equal to 1/88 of the total energy communicated.

But if a piano is to represent a fairly good analogy of the Jeans' cube, it must have many more keys extending beyond any limit to the right into the ultrasonic region (Fig. 4). Thus the energy communicated to one string in an audible region would travel to the right into the region of higher pitches and be lost in the

---

pected because of virtual electron-pair formation. But Jeans and Planck did not know this.

Fig. 4. A piano with an unlimited number of keys extending into the ultrasonic region all the way to infinite frequencies. The equipartition law would require all the energy supplied by a musician to one of the low-frequency keys to travel all the way into the ultrasonic region out of the audible range!

infinitely far regions of the ultrasonic vibrations, and a piece of music played on such a piano would turn into a sharp shrill. Similarly *the energy of red light injected into Jeans' cube would turn into blue, violet, ultraviolet, X-rays, γ-rays, and so on without any limit.* It would be foolhardy to sit in front of a fireplace since the red light coming from the friendly glowing cinders would quickly turn into dangerous high-frequency radiation of fission products!

The runaway of energy into the high-pitch region does not represent any real danger to concert pianists, not only because the keyboard is limited on the right, but mostly because, as was mentioned before, the vibration of each string is damped too fast to permit a transfer of even a small part of energy to a neighboring string. In the case of radiant energy, however, the situation is much more serious, and, if the Equipartition Law should hold in that case, the open door of a boiler would be an excellent source of X- and γ-rays. Clearly something must be wrong with the arguments of nineteenth-century physics, and some drastic changes must be made to avoid the Ultraviolet Catastrophe, which is expected theoretically but never occurs in reality.

MAX PLANCK AND THE QUANTUM OF ENERGY

The problem of radiation-thermodynamics was solved by Max Planck, who was a 100 per cent classical physicist (for which he cannot be blamed). It was he who originated what is now known as *modern physics*. At the turn of the century, at the December 14, 1900 meeting of the German Physical Society, Planck presented his ideas on the subject, which were so unusual and so grotesque that he himself could hardly believe them, even though they caused intense excitement in the audience and in the entire world of physics.

Max Planck was born in Kiel, in 1858, and later moved with his family to Munich. He attended Maximilian Gymnasium (high school) in Munich and, after graduation, entered the University of Munich, where he studied physics for three years. The following year he spent at the University of Berlin, where he came in contact with the great physicists of that time, Herman von Helmholtz, Gustav Kirchhoff, and Rudolph Clausius, and learned much about the theory of heat, technically known as thermodynamics. Returning to Munich, he presented a doctoral thèsis on the Second Law of Thermodynamics, receiving his Ph.D. degree in 1879, and then became an instructor at that university. Six years later he accepted the position of associate professor in Kiel. In 1889 he moved to the University of Berlin as an associate professor, becoming a full professor in 1892. The latter position was, at that time, the highest academic position in Germany, and Planck kept it until his retirement at the age of seventy. After retirement he continued his activities and delivered public speeches until his death at the age of almost ninety. Two of his last papers (*A Scientific Autobiography* and *The Notion of Causality in Physics*) were published in 1947, the year he died.

Planck was a typical German professor of his time, serious and probably pedantic, but not without a warm human feeling, which is evidenced in his correspondence with Arnold Sommerfeld who, following the work of Niels Bohr, was applying the Quantum Theory to the structure of the atom. Referring to the quantum as Planck's notion, Sommerfeld in a letter to him wrote:

> *You* cultivate the virgin soil,
> Where picking flowers was *my* only toil.

and to this answered Planck:

> *You* picked flowers—well, so have *I*.
> Let them be, then, combined;
> Let us exchange our flowers fair,
> And in the brightest wreath them bind.§

For his scientific achievements Max Planck received many academic honors. He became a member of the Prussian Academy of Sciences in 1894, and was elected a foreign member of the Royal Society of London in 1926. Although he made no contribution to the science of astronomy, one of the newly discovered asteroids was called Planckiana in his honor.

Throughout all his long life Max Planck was interested almost exclusively in the problems of thermodynamics, and the many papers he published were im-

§ *Scientific Autobiography* by M. Planck. Translated by F. Gaynor. New York: Philosophical Library (1949).

portant enough to earn him the honorable position of full professor in Berlin at the age of thirty-four. But the real outburst in his scientific work, the discovery of the *quantum of energy,* for which, in 1918, he was awarded the Nobel Prize, came rather late in life, at the age of forty-two. Forty-two years is not so late in the life of a man in the usual run of occupations or professions, but it usually happens that the most important work of a theoretical physicist is done at the age of about twenty-five, when he has had time to learn enough of the existing theories but while his mind is still agile enough to conceive new, bold revolutionary ideas. For example, Isaac Newton conceived the Law of Universal Gravity at the age of twenty-three; Albert Einstein created his Theory of Relativity at the age of twenty-six; and Niels Bohr published his Theory of the Atomic Structure at the age of twenty-seven. In his small way, the author of this book also published his most important work, on natural and artificial trans- formations of the atomic nucleus, when he was twenty- four. In his lecture Planck stated that according to his rather complicated calculations the paradoxical con- clusions obtained by Rayleigh and Jeans could be remedied and the danger of the Ultraviolet Catastro- phe avoided *if one postulates that the energy of elec- tromagnetic waves (including light waves) can exist only in the form of certain discrete packages, or quanta, the energy content of each package being directly pro- portional to the corresponding frequency.*

Theoretical considerations in the field of statistical physics are notoriously difficult, but by inspecting the graph in Fig. 5 one can get some notion of how Planck's postulate "discourages" radiant energy from leaking into the limitless high-frequency region of the spectrum.

In this graph the frequencies possible within a "one-

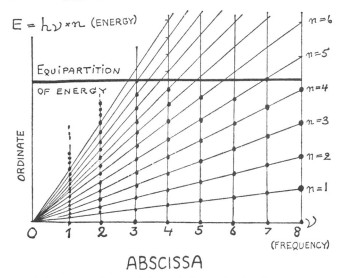

Fig. 5. *If, according to Planck's hypothesis, the energy corresponding to each frequency ν must be an integer of the quantity hν, the situation is quite different from that shown in the previous diagram. For example, for ν = 4 there are eight possible vibration states, whereas for ν = 8 there are only four. This restriction reduces the number of possible vibrations at high frequencies and cancels Jeans' paradox.*

dimensional" Jeans' cube are plotted on the abscissa axes and marked 1, 2, 3, 4, etc.; on the ordinate axes are plotted the vibration energies that can be allotted to each possible frequency. According to classical physics any value of energy (that is, any point on the vertical lines drawn through 1, 2, 3, etc.) is permitted, the distribution resulting statistically in the Equipartition of Energy among all possible frequencies. On the other hand, Planck's postulate permits only a discrete set of energy values, equal to one, two, three, etc., energy packages corresponding to the given frequency. Since the energy contained in each package is assumed to be proportional to the frequency, we obtain the permitted energy values shown by large black dots in the diagram.

The higher the frequency, the smaller is the number of possible energy values below any given limit, a fact which restricts the capacity of the high-frequency vibrations to take up more additional energy. As a result, the amount of energy that can be taken by high-frequency vibrations becomes finite in spite of their infinite number, and everything is dandy.

It has been said that there are "lies, white lies, and statistics," but in the case of Planck's calculations the statistics turned out to be well-nigh true. He had obtained for energy distribution in thermal radiation spectrum a theoretical formula that stood in perfect agreement with the observation shown in Fig. 2.

While the Rayleigh-Jeans formula shoots sky high, demanding an infinite amount of total energy, Planck's formula comes down at high frequencies and its shape stands in perfect agreement with the observed curves. Planck's assumption that the energy content of a radiation quantum is proportional to the frequency can be written as:

$$E = h\nu$$

where ν (the Greek letter nu) is the frequency and h is a universal constant known as *Planck's Constant,* or the *quantum constant.* In order to make Planck's theoretical curves agree with the observed ones, one has to ascribe to h a certain numerical value, which is found to be 6.77×10^{-27} in the centimeter-gram-second unit system.¶

The numerical smallness of that value makes quantum theory of no importance for the large-scale phenomena which we encounter in everyday life, and it

¶ The physical dimension of the quantum constant h is a product of energy and time, or /erg · sec/ in c.g.s. units, and is known in classical mechanics as *action;* action appears in many important considerations, such as the Hamilton's Principle of Least Action.

emerges only in the study of the processes occurring on the atomic scale.

LIGHT QUANTA AND THE PHOTOELECTRIC EFFECT

Having let the spirit of quantum out of the bottle, Max Planck was himself scared to death of it and preferred to believe the packages of energy arise not from the properties of the light waves themselves but rather from the internal properties of atoms which can emit and absorb radiation only in certain discrete quantities. Radiation is like butter, which can be bought or returned to the grocery store only in quarter-pound pack-

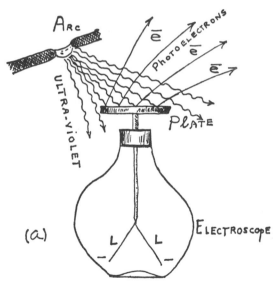

Fig. 6. Experimental Studies of the Photoelectric Effect. In (a) a primitive method for a demonstration of the photoelectric effect is illustrated. The ultraviolet radiation emitted by an electric arc ejects the electrons from a metal plate attached to an electroscope. The negatively charged leaves L, which have been repelling each other, lose charge and collapse. In (b) the modern method is shown. Ultraviolet

ages, although the butter as such can exist in any desired amount (not less, though, than one molecule!).
Only five years after the original Planck proposal, the
light quantum was established as a physical entity existing independently of the mechanism of its emission or
absorption by atoms. This step was taken by Albert
Einstein in an article published in 1905, the year of
his first article on the Theory of Relativity. Einstein
indicated that the existence of light quanta rushing
freely through space represents a necessary condition
for explaining empirical laws of the photoelectric effect; that is, the emission of electrons from the metallic
surfaces irradiated by violet or ultraviolet rays.

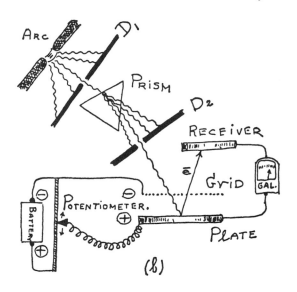

*radiation from an electric arc passes through a prism
allowing only one selected frequency to fall on the plate.
Turning the prism, one can select a monochromatic light
and direct it to the plate. The energy of photoelectrons is
measured by their ability to get through from plate to
receiver, moving against the electric force produced by a
potentiometer between plate and grid.*

An elementary arrangement for demonstrating photoelectric effect, shown in Fig. 6a, consists of a negatively charged ordinary electroscope with a clean metal plate *P* attached to it. When a light from an electric arc *A*, which is rich in violet and ultraviolet rays, falls on the plate, one observes that the leaves *L* of the electroscope collapse as the electroscope discharges. That negative particles (electrons) are discharged from the metal plate was demonstrated repeatedly, by the American physicist Robert Millikan (1868–1953) among others. If a glass plate, which absorbs ultraviolet radiation, is interposed between the arc and the metal plate, electrons are not given off, conclusive evidence that the action of the rays causes the emission. A more elaborate arrangement used for the detailed study of the laws of photoelectric effect is shown schematically in Fig. 6b. It consists of:

1. A quartz or fluoride prism (transparent for ultraviolet) and a slit permitting the selection of a monochromatic radiation of desired wavelength.

2. A set of rotating discs with triangular openings of various sizes, permitting a change in the intensity of radiation.

3. An evacuated container somewhat similar to the electron tubes used in radio sets. A variable electric potential is applied between the plate *P*, from which the photoelectrons are emitted, and the grid *G*. If the grid is charged negatively, and potential difference between the grid and the plate is equal to or larger than the kinetic energy of photoelectrons expressed in electron volts, no current will flow through the system. In the opposite case there will be a current, and its strength can be measured by the galvanometer *GM*. Using this arrangement, one can measure the number and the kinetic energy of electrons ejected by the incident light of any given intensity and wavelength (or frequency).

The study of photoelectric effect in different metals resulted in two simple laws:

I. *For light of a given frequency but varying intensity the energy of photoelectrons remains constant while their number increases in direct proportion to the intensity of light* (Fig. 7a).

II. *For varying frequency of light no photoelectrons are emitted until that frequency exceeds a certain limit ν_0, which is different for different metals. Beyond that frequency threshold the energy of photoelectrons increases linearly, being proportional to the difference between the frequency of the incident light and the critical frequency ν_0 of the metal* (Fig. 7b).

These well-established facts could not be explained on the basis of the classical theory of light; in some points they even contradicted it. Light is known to be short electromagnetic waves, and the increase of intensity of light must mean an increase of the oscillating electric and magnetic forces propagating through space. Since the electrons apparently are ejected from the metal by the action of electric force, their energy should increase with the increase of light intensity, instead of remaining constant as it does. Also, in the classical electromagnetic theory of light, there was no reason to expect a linear dependence of the energy of photoelectrons on the frequency of the incident light.

Using Planck's idea of light quanta and assuming the reality of their existence as independent energy packages flying through space, Einstein was able to give a perfect explanation of both empirical laws of photoelectric effect. He visualized the elementary act of the photoelectric effect as the result of a collision between a single incident light quantum and one of the conductivity electrons carrying electric current in the metal. In this collision the light quantum vanishes, giving its entire energy to the conductivity electron at the metallic surface. But, in order to cross the surface and

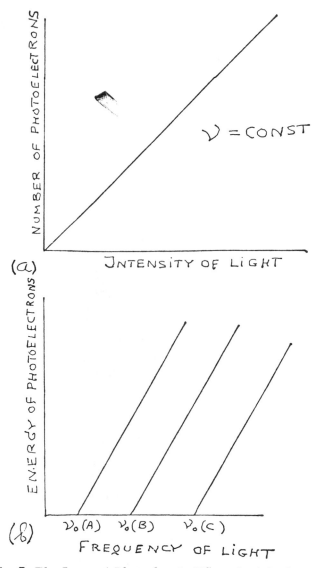

Fig. 7. The Laws of Photoelectric Effect. In (a) the number of electrons is plotted as the function of the intensity of
the incident monochromatic light. In (b) the energy of
photoelectrons is shown as the function of the frequency of
the incident monochromatic light for three different metals:
A, B and C.

to get into the free space, the electron must spend a certain amount of energy disengaging itself from the attraction of metallic ions. This energy, known by the somewhat misleading name of "work function," is different for different metals and is usually denoted by a symbol W. Thus, the kinetic energy K with which a photoelectron gets out of the metal is:

$$K = h(v - v_0) = hv - W$$

where v_0 is the critical frequency of light below which the photoelectric effect does not occur. This picture explains at once the two laws derived from experiment. If the frequency of the incident light is kept constant, the energy content of each quantum remains the same, and the increase of light intensity results only in the corresponding increase of the number of light quanta. Thus more photoelectrons are ejected, each of them with the same energy as before. The formula giving K as the function of v explains the empirical graphs shown in Fig. 7b, predicting that the slope of the line should be the same for all metals having a numerical value equal to h. This consequence of Einstein's picture of photoelectric effect stands in complete agreement with experiment and leaves no doubt of the reality of light quanta.

THE COMPTON EFFECT

An important experiment proving the reality of light quanta was performed in 1923 by an American physicist, *Arthur Compton,* who wanted to study a collision of light quantum with an electron moving freely through space. The ideal situation would be to observe such collisions by sending a beam of light through an electron beam. Unfortunately, the number of electrons in even the most intense electron beams available is so small that one would have to wait for centuries for

a single collision. Compton solved the difficulty by using X-rays, the quanta of which carry very large
amounts of energy because of the very high frequency
involved. As compared with the energy carried by each
X-ray quantum, the energy with which electrons are
bound in the atoms of light elements can be disregarded
and one can regard them (the electrons) as being un-
bound and quite free. Considering a free collision between light quantum and an electron in the same way
as one considers a collision between two elastic balls,
one would expect that the energy, and hence the frequency, of scattered X-rays would decrease with the
increasing scattering angle. Compton's experiments
(Fig. 8) stood in complete agreement with this theoretical prediction, and with the formula derived on the
basis of conservation of energy and mechanical momentum in the collision of two elastic spheres. This
agreement gave additional confirmation of the existence of light quanta.

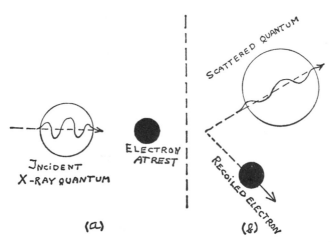

*Fig. 8. Compton scattering of X-rays. Notice that after the
collision the wavelength of X-ray quantum increases because of loss of energy to the electron.*

CHAPTER II

N. BOHR AND QUANTUM ORBITS

The discovery that light propagates through space and can be emitted or absorbed by matter only in the form of discrete energy packages (light quanta) whose energy content is strictly defined by their vibration frequencies, had a profound influence on the contemporary views concerning the structure of atoms themselves. When, in 1897, J. J. Thomson proved by direct experiments that tiny negatively charged particles (electrons) can be extracted from atoms, leaving behind positively charged residues (ions), it became clear that atoms are not, as the Greek meaning of their name implies, indivisible constituent units of matter, but, quite to the contrary, rather complex systems formed by positively and negatively charged parts. Thomson visualized the atom as being formed by some positively charged substance distributed more or less uniformly through the entire body, with negatively charged electrons imbedded in it as are raisins inside a round loaf of raisin bread. The electrons are attracted to the center of the positive charge distribution and repelled by one another according to Coulomb's law of electric interactions, and the normal state of the atom is attained when these two opposing sets of forces are in equilibrium. If an atom is disturbed (or, as physicists say, "excited") by a collision with another atom or a passing free electron, its inner electrons (like strings in a grand piano) begin to vibrate around their equilibrium positions and emit a set of characteristic light frequencies, which should account for the observed line spectra. The atoms of different chemical elements possess different numbers of differently distributed inner electrons with different characteristic frequencies, and thus differ in their observed optical spectra (Fig. 9). If Thomson's model of an atom was accepted, it was possible, by the methods of classical mechanics, to calculate the equilibrium distribution of electrons within the body of an atom containing a given amount of inner electrons, and it was expected that the sets of calculated characteristic vibration frequencies would coincide with the observed line spectra of various elements.

Thomson himself and his students carried out complicated computations to find the configurations of interatomic electrons for which the calculated vibration frequencies should coincide with the observed frequencies in the line spectra of various chemical elements. The results were disappointingly negative. The theoretically calculated spectra based on Thomson's model

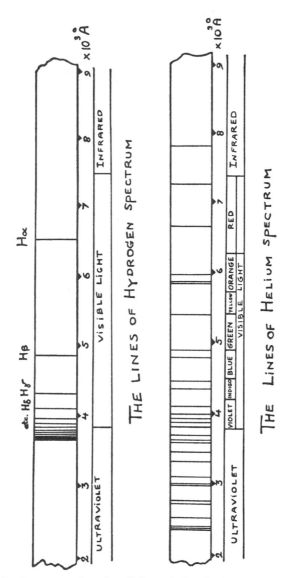

Fig. 9. Compare the simplicity of the hydrogen spectrum, produced by the motion of only one electron, with the complexity and seeming lack of order caused by the two electrons in the spectrum of helium. Both spectra continue much farther into ultraviolet and infrared regions.

looked not at all like the observed spectra of any of the chemical elements. It became more and more evident that some revolutionary change should be made in Thomson's classical model of the atom. This point was especially stressed by a young Danish physicist, Niels Bohr, who, after receiving his Ph.D. from the University of Copenhagen for a paper on the theory of the passage of charged particles through matter, arrived in 1911 at the Cavendish Laboratory of Cambridge University, in England, to join the group working under its director, J. J. Thomson. Bohr argued that since light must no longer be treated as continuously propagating waves but instead be endowed with the mysterious additional properties of emission and absorption as discrete energy packages of well-defined size, the classical Newtonian mechanics on which Thomson's atomic model was based should be changed correspondingly. If the electromagnetic energy of light is "quantized"—that is, restricted to definite portions of one, two, three, or more light quanta ($h\nu$, $2h\nu$, $3h\nu$, etc.)—isn't it reasonable to assume that the mechanical energy of atomic electrons is quantized, too, that it can assume only a discrete set of values, the intermediate values being prohibited by some yet undiscovered law of nature? Indeed, it would be odd if atomic systems built according to the laws of classical Newtonian mechanics, as Thomson's atomic model was, should emit and absorb light in the form of Planck's light quanta, which do not fit at all into the frame of classical physics!

RUTHERFORD'S THEORY OF THE NUCLEAR ATOM

J. J. Thomson did not like these revolutionary ideas of the young Dane. A number of sharp arguments forced Bohr to decide to abandon Cambridge and

Niels and Mrs. Bohr enjoying a motorcycle ride.
(Photographed by the author)

spend the rest of his foreign fellowship at some place where his still vague ideas about the quantization of the electron's motion in atoms would meet less opposition. His choice was the University of Manchester, where the chair of physics was held by a New Zealand farmer's son and former student of Thomson's by the name of Ernest Rutherford, who later received the title of Sir Ernest and then Lord Rutherford of Nelson for his scientific discoveries. When Bohr arrived in Manchester, Rutherford was in the midst of his epoch-making studies of the internal structure of atoms by shooting them with high-energy projectiles known as "α-particles" which were emitted by the then newly discovered radioactive elements. In earlier studies, mostly carried out at McGill University, in Canada, Rutherford had been able to prove that α-particles emitted by radioactive elements are nothing other than positively charged atoms of helium moving with tremendously high velocities never before encountered in physics. Emission of the α-particles from unstable heavy atoms of radioactive elements was often followed by the emission of electrons (β-particles) and high-frequency electromagnetic radiation (γ-rays) similar to ordinary X-rays but having much shorter wavelength. If one wants to crack something, one naturally chooses as a projectile a solid iron ball rather than a light ping-pong ball, and Rutherford figured that massive α-particles would penetrate much more easily into the atomic interior than light β-particles. The arrangement was quite simple (Fig. 10). A small amount of radioactive material, Ra for example, emitting α-particles was put at a pinhead and placed at a certain distance from a thin foil (F) made from the metal to be investigated. A thin beam of α-particles was formed after the ray passed a diaphragm (D). In passing through the foil, α-particles collided with the constituent atoms and a fraction

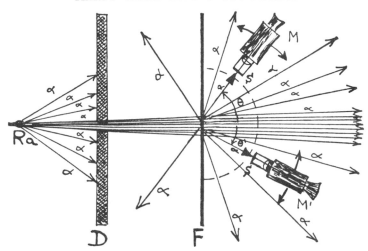

Fig. 10. Rutherford apparatus for studying angular dependence of α-scattering.

of the particles were scattered in different directions on the opposite side of the foil. Falling on the fluorescent screen (*S*) placed behind the foil, each α-particle produced a little spark (scintillation) at the point of impact. Observing these scintillations through a microscope (*M*), one could count the number of particles scattered at different angles from the original direction, just as in target shooting with firearms one can measure the deviation of the bullet holes from the bull's-eye. In his experiments, Rutherford noticed that, whereas the majority of the particles passed through the foil almost without deflection, forming a luminous spot (bull's-eye) opposite the diaphragm opening, some were scattered at a quite considerable angle. Somewhat different experimental arrangements have shown that in a few cases α-particles are thrown almost backward toward the source.

This observation stood in direct contradiction of what would be expected on the basis of Thomson's atomic model. By passing through the atom, the incident α-particle can be deflected from its original track either by the electric attraction of interatomic electrons, or by the electric repulsion of the spread-out positive charge. Interaction with the electrons, which are almost 10,000 times lighter than α-particles, certainly could not produce any noticeable deflection in α-particle motion. On the other hand, the positively charged material in Thomson's model was distributed too thinly through the entire body of the atom to be able to cause any appreciable deflection of the α-particles passing through it. Indeed, if we throw an iron ball at a piece of coal it will bounce off it at an odd angle, perhaps breaking the coal into several pieces. But, if we grind the same piece of coal into fine powder and throw the same ball through the resulting coal dust cloud, it will pass through without any deflection. The observed very large deflections in Rutherford's scattering experiments definitely proved that the positive charge (associated with most of the mass) of the atom is not distributed all through its body, as in the previous example of coal dust cloud, but is concentrated, like the solid piece of coal, in a little hard nut—the nucleus. The experimentally observed dependence of the number of α-particles scattered at various directions on the angle of scattering stood in perfect agreement with the theoretical formula for the scattering of particles moving in a field of a repulsive central force whose strength is inversely proportional to the square of the distance.

Thus was born Rutherford's atomic model. With its light negatively charged electrons moving through free space around a positively charged heavy nucleus in the center, it somewhat resembled the Solar System. Since Coulomb's law of electric attraction is mathematically identical with Newton's law of gravity (both forces being inversely proportional to the square of the distance), atomic electrons move around the nucleus along the circular or elliptic orbits, just as do planets around the Sun.

But there is one great difference which lies in the fact that, whereas the Sun and the planets are electrically neutral, the atomic nucleus and electrons carry heavy electric charges. It is well known that the oscillating electric charges produce diverging electromagnetic waves. Rutherford's atomic model can be considered a miniature broadcasting station operating on ultra-high frequency. Using the classical theory of electromagnetic emission, one easily calculates that light waves emitted by electrons circling the atomic nucleus will take away into space all the electrons' energy within about one hundred-millionth of a second. Having lost all their energy, atomic electrons must fall into the nucleus and the atom cease to exist!

Strictly speaking, similar losses of energy are expected also in the case of planets of the Solar System. According to Einstein's General Theory of Relativity, the oscillation of gravitating masses also emits so-called "gravitational waves" which take away energy. But, because of the small value of Newton's constant, planetary energy losses through gravitational emission are extremely small, and, since their formation some four or five billion years ago, the planets cannot have lost more than a few per cent of their original energy.

QUANTIZING A MECHANICAL SYSTEM

But what to do about the atoms built according to Rutherford's model? Theoretically, as we have said, they cannot exist longer than one hundred-millionth of a second, but in reality they do exist for eternity. This was the question which confronted young Bohr upon his arrival in Manchester.

Staggering contradictions of this kind between theoretical expectations on the one side and observational facts or even common sense on the other are the main factors in the development of science. A. A. Michelson's failure to detect the motion of the Earth through a luminiferous ether† led Einstein to the formulation of the Theory of Relativity, which altered our common sense notions about space and time and caused profound changes in classical physics. Similarly, the Ultraviolet Catastrophe, discussed in the previous chapter, led Planck to a completely novel idea of light quanta.

The theoretical impossibility of the experimentally proved Rutherford model of an atom resonated with Bohr's hidden feeling that, if the electromagnetic energy is quantized, mechanical energy must be quantized too, even though perhaps in a somewhat different way. In fact, when an excited atom is emitting light quantum with the energy $h\nu$, its mechanical energy must decrease exactly by this amount. Since the atomic spectra consist of a series of discrete, sharply defined lines, the energy differences between various possible states of an atom must also have sharply defined values, and so must the absolute energies of these states themselves. This leads to the idea that the atomic mechanism is somewhat similar to an automobile gear box. One can make it run in the 1st (bottom) gear, 2nd gear, 3rd gear, etc., but never in 1½th gear or 3⅗th gear. . . .

Let E_1, E_2, E_3, E_4, etc., be the possible energy values of different states of an atom arranged in increasing order (Fig. 11). An atom always has some internal energy, but when this energy has fallen to the lowest possible level, E_1, none of it is available for emission of a light quantum. This E_1 level is the *normal* or

† See Bernard Jaffe, *Michelson and the Speed of Light.* Doubleday, Science Study Series (1960).

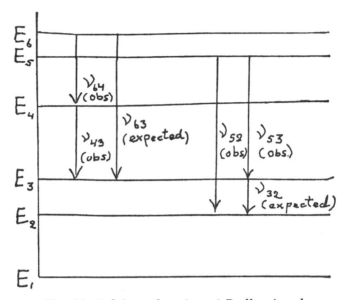

Fig. 11. Bohr's explanation of Rydberg's rule.

ground state of the atom, in which it can exist for eternity. It is known as the *zero point* energy, and in the case of an oscillator is $1/2h\nu$. Suppose now that the atom is brought into an excited state with some higher energy E_n. The energy can be achieved, for example, by subjecting the gas to a very high temperature, as in the atmosphere of the Sun, where the atoms are brought to excited states by violent thermal collisions among themselves. Another way to excite atoms is to pass a high-tension electric discharge through a glass tube filled with rarefied gas.‡ The atoms are excited

‡ One must use rarefied gas in order to give the electrons sufficiently long time intervals between the collisions to regain energy lost in each collision by being accelerated in the applied electric field. Under normal atmospheric pressure gases do not conduct electricity, and once the tension becomes very high a sudden breakdown occurs in the form of a spark.

by the impact of fast electrons rushing through the tube from the negative electrode (cathode) to the positive one (anode). Such gadgets, in which the gas becomes luminous on passage of the high-tension electric discharge, were originally known as Geissler tubes after their inventor, Heinrich Geissler. You see them everywhere today in luminous street signs and other lighting devices.

When an atom is excited to the mth energy state E_m, it can return to some lower energy state, E_n ($n < m$) by liberating the balance of energy in the form of a light quantum. Thus we can write

$$h\nu_{m,n} = E_m - E_n$$

or

$$\nu_{m,n} = \frac{E_m - E_n}{h}$$

The two indices at $\nu_{m,n}$ indicate that this particular frequency in the spectrum corresponds to the transition from the mth quantum state of motion to the nth quantum state.

This picture of the light quanta emission resulting from the transition of the atom from a higher energy state to a lower one has a very interesting consequence. Suppose in the spectrum of some element we observe two lines corresponding to the transition from the 6th quantum state to the 4th and from the 4th to the 3rd (left side in Fig. 11). Then it is possible that the transition can occur straight from the 6th state to the 3rd, and we find a line with a frequency

$$\nu_{6,3} = \nu_{6,4} + \nu_{4,3}$$

The situation shown on the right side of the same figure is the opposite one. From the fact that we observe the frequencies $\nu_{5,2}$ and $\nu_{5,3}$ it follows that we may also observe the frequency

$$\nu_{3,2} = \nu_{5,3} - \nu_{5,2}$$

The Swiss spectroscopist W. Ritz discovered this law of *Addition and Subtraction* when Niels Bohr was still a school boy. But in pre-quantum spectroscopy Ritz's rule and other similar numerical regularities between observed frequencies were only puzzling riddles that could not be explained in a reasonable way. They became very helpful, however, for Niels Bohr, in his attempt to solve the problem of light emission and absorption of atoms by introducing the idea of the discrete quantum states of atomic electrons.

For his first studies Bohr selected the atom of hydrogen, the lightest and presumably the most simply constructed atom, which was also known to possess a very simple spectrum. In 1885 a Swiss schoolteacher, J. J. Balmer, who was interested in regularities of atomic line spectra, discovered that the frequency of the visible part of hydrogen can be represented with great precision by a very simple formula. The frequencies of these lines, shown in Fig. 9 top (where they are plotted in terms of wavelength $\lambda = c/\nu$), are given in the following table:

$$H_\alpha \text{———} \nu_1 = 4.569 \times 10^{14} \sec^{-1}\text{§}$$
$$H_\beta \text{———} \nu_2 = 6.168 \times 10^{14} \sec^{-1}$$
$$H_\gamma \text{———} \nu_3 = 6.908 \times 10^{14} \sec^{-1}$$
$$H_\delta \text{———} \nu_4 = 7.310 \times 10^{14} \sec^{-1}$$

As the reader can himself verify, these figures can be obtained from the formula

$$\nu_{m,n} = 3.289 \times 10^{15}\left(\frac{1}{4} - \frac{1}{m^2}\right)\sec^{-1}$$

where m assumed the values: 3, 4, 5, 6.¶ For larger n's

§ Sec^{-1} means "per second"; cm^{-1} means "per centimeter"; apples \$$^{-1}$ means "apples per dollar."
¶ The numerical coefficient in the above formula is usually denoted by R and known as the Rydberg constant, although it should be more properly called the Balmer constant.

the frequencies fall into the ultraviolet region, and the lines become more and more crowded, converging to the value

$$3.289 \times 10^{15} \times \frac{1}{4} = 6.225 \times 10^{14} \sec^{-1}$$

In Bohr's picture of the relation between the emitted light quanta $h\nu_{m,n}$ and the energy states E_m and E_n (or levels) of the atom, the Balmer formula tells us that the mth line of the series is due to the transition from the mth state of an excited hydrogen atom to the second state (since $4 = 2^2$). If instead of

$$\frac{1}{4} = \frac{1}{2^2}$$

in Balmer's formula one substitutes

$$1 = \frac{1}{1^2}$$

and puts $m = 2, 3, 4$, etc., one obtains a sequence of lines which fall into the far ultraviolet region and were actually discovered by Theodore Lyman. If, on the other hand, one chooses for the first term in the Balmer

$$\frac{1}{9} = \frac{1}{3^2}, \text{ or } \frac{1}{16} = \frac{1}{4^2},$$

formula one obtains the light frequencies which fall into the far infrared region and were found by Friedrich Paschen and Frederick Brackett, respectively. Thus the mechanical quantum states must look as is shown in Fig. 12, which also indicates the transitions resulting in the emission of Lyman, Balmer, Paschen, and Brackett series.

Thus each line in the entire spectrum is characterized by two indices, m and n, of the two quantum levels, between which the transition occurs (starting

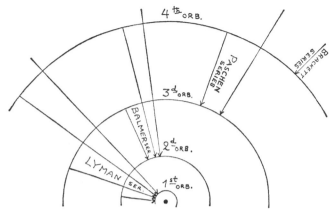

*Fig. 12. Bohr's original model of the hydrogen atom. Ly-
man series are in the ultraviolet, Balmer series in the visible
part of the spectrum. Paschen and Brackett series are in
infrared.*

from mth and finishing on nth). Since the energy of
the light quantum is equal to the energy difference between the state from which it started and the state at
which it ended, the generalized Balmer formula should
be rewritten as:

$$h\nu_{m,n} = Rh\left[\left(-\frac{1}{m^2}\right) - \left(-\frac{1}{n^2}\right)\right]$$

or

$$h\nu_{m,n} = \left(-\frac{Rh}{m^2}\right) - \left(-\frac{Rh}{n^2}\right)$$

where the two quantities in the parentheses represent
the energy levels E_m and E_n. The reason for writing
these energies as negative quantities is that, conventionally, one ascribes zero energy to the state of the
system when all its parts are at an infinite distance from
one another. Thus, if the energy of the system is positive, it will not hold together, and all its components
will fly apart. In a stable system like that of the planets

revolving around the Sun, or of electrons revolving around the atomic nucleus, the energy is negative, and it would require a supply of energy from the outside to take them apart.

To explain the values of energy of the different states of the hydrogen atom as given by the above formula, Bohr made two simplifying assumptions:

1st: That the atom of hydrogen, being the simplest atom of the entire periodic system of elements, contains *only one* electron.

2nd: That different quantum states of the hydrogen atom correspond to the motion of that electron along the *circular orbits* with different radii. Making these assumptions, one should be able to find the quantum orbits of the electrons from the relation:

$$E_n = -\frac{Rh}{n^2}$$

Consider the orbital motion of an electron in the nth excited state of the hydrogen atom, and write r_n and v_n for the radius and the orbital velocity of the electron in the nth state. The electron's mass is m_e, and its charge $-e$, whereas the charge of the nucleus (proton in this case) is $+e$. The condition of circular motion of the electron is that the electrostatic attraction force $-\dfrac{e^2}{r^2}$ is balanced by the centrifugal force $+\dfrac{mv^2}{r}$. From

$$-\frac{e^2}{r^2} + \frac{m_e v^2}{r} = 0$$

follows

$$v = \frac{e}{\sqrt{m_e r}}$$

giving the velocity v of the electron necessary for its motion along the circle of the radius r. According to this equation of classical mechanics an electron can move along *any* circular orbit provided it has the necessary velocity.

What is the quantum condition which selects only the orbits with the energies $E_n = -Rh/n^2$?

In the Quantum Theory of Radiation described in the previous section we stated that the vibration of a given frequency ν can carry only the energy of one, two, three, or more light quanta so that $E_n = nh\nu$ ($n = 1, 2, 3$, etc.). We can rewrite it in the form:

$$\frac{E_n}{\nu} = -nh$$

meaning that the quantity E/ν can be only an integer multiple of the quantum constant h. It may be mentioned here that the physical dimension of h is:

$$|\text{action}| = \frac{|\text{energy}|}{|\text{frequency}|} = \frac{|\text{mass}| \cdot |\text{velocity}|^2}{|\text{frequency}|} =$$

$$= \frac{|\text{mass}| \, |\text{length}|^2}{|\text{time}|^{-1} \cdot |\text{time}|^2} = |\text{mass}| \cdot \frac{|\text{length}|}{|\text{time}|} \cdot |\text{length}| =$$

$$= |\text{mass}| \cdot |\text{velocity}| \cdot |\text{length}|.$$

The product of the *mass* of a particle by its *velocity* and by the *distance* it travels is a well-known quantity called the *action,* and plays an important part in classical analytical mechanics. For example, "the principle of least action" formulated by a French mathematician, P. L. M. de Maupertuis in 1747, states that a particle subjected to mechanical forces will travel from point A to point B along a trajectory for which the "total action" from A to B will be either the smallest or the largest of all other possible trajectories between these two points. Planck's Law of Light Quanta adds to Maupertius's principle a supplementary condition that:

The total action must always be an integer multiple of h.

In the case of the closed circular trajectory of an electron around the nucleus, the quantum condition will demand that the product of the electron mass, its velocity, and the distance covered in one revolution must be an integer multiple of h. Thus, for the nth Bohr's orbit:

$$m_e \cdot v_n \cdot 2\pi r_n = nh$$

$$m_e \cdot \frac{e}{\sqrt{m_e r_n}} \cdot 2\pi r_n =$$

$$= 2\pi e \sqrt{m_e} \sqrt{r_n} = nh$$

or

$$r_n = \frac{h^2}{4\pi^2 e^2 m} \cdot n^2$$

We shall now calculate the total energy E_n of an electron on the nth orbit, which is the sum of its kinetic energy K and potential energy U. Using the expression for velocity $v = e/\sqrt{m_e r}$ given earlier, and remembering that the potential energy of two charges $+e$ and $-e$ located at the distance r apart is $+e^2/r$, we write:

$$E_n = K_n + U_n = \frac{1}{2} m_e \frac{e^2}{m_e r_n} + \frac{e^2}{r_n} =$$

$$= \frac{1}{2} \frac{e^2}{r_n} - \frac{e^2}{r_n} = -\frac{1}{2} \frac{e^2}{r_n}$$

Substituting into that the expression for r_n from the earlier formula, we obtain

$$E_n = -\frac{4\pi^2 e^4 m_e}{h^2} \cdot \frac{1}{n^2}$$

which coincides with the empirical expression

$$E_n = -\frac{Rh}{n^2}$$

obtained from the Balmer formula if we put:

$$R = \frac{4\pi^2 e^4 m_e}{h^3}$$

When Bohr substituted into this expression the numerical values e, m_e, and h, he obtained $R = 3.289 \times 10^{15}$ sec^{-1}, which is exactly its empirical value obtained by spectroscopic observation. Thus, the problem of quantization of a mechanical system was successfully solved.

SOMMERFELD'S ELLIPTICAL ORBITS

The original Bohr paper on the hydrogen atom was followed soon by that of a German physicist, Arnold Sommerfeld, who extended Bohr's ideas to the case of elliptical orbits. The motion of a particle in the field of central force is characterized in general by two (polar) coordinates, its distance r from the center of attraction and its positional angle (azimuth) ϕ in respect to the major axis of the ellipse as indicated in the figure (Fig. 13); r has the maximum value when $\phi = 0$, decreases to its minimum value at $\phi = \pi$, and increases again to its maximum value at $\phi = 2\pi$. Thus, in contrast to Bohr's circular orbits where r remains constant and only ϕ changes, the motion along Sommerfeld's elliptical orbits is characterized by two independent coordinates, r and ϕ. It follows that each quantized elliptical orbit must be characterized by *two* quantum numbers: azimuthal quantum number n_ϕ and the radial quantum number n_r. Applying Bohr's quantum conditions that the total mechanical actions for *azimuthal* and *radial* components of motion must be integer numbers n_ϕ and n_r of h, Sommerfeld obtained for the energy of the quantized elliptical motion the formula:

$$E_{n_\phi, n_r} = -\frac{Rh}{(n_\phi + n_r)^2}$$

This is exactly the same as Bohr's original formula except that, instead of the square of an integer, the denominator is the square of a sum of two arbitrary integers which is, of course, an arbitrary integer itself. Putting $n_r = 0$ we get, as a special case, Bohr's circular orbits. If $n_r \neq 0$ we get elliptical orbits with different degrees of ellipticity. But the energies of all orbits corresponding to the same sum $n_\phi + n_r$ is exactly the

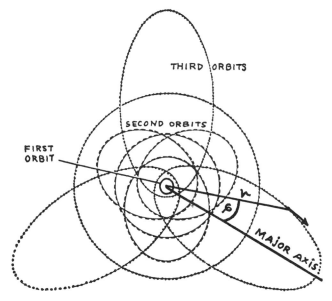

Fig. 13. Circular and elliptical quantum orbits in the hydrogen atom. The first circular orbit (solid line) corresponds to the lowest energy of the electron. The next four orbits, one circular and three elliptical (dashed lines) correspond to the same energy, which is higher than on the first orbit. The next nine orbits (dotted lines), only four of which are shown in the figure, correspond to still higher energy (the same for all nine).

same in spite of their different shapes. The sum n_ϕ + n_r, usually denoted simply by n, is known as the *principal quantum number.*

It may be remarked here that relativistic treatment of the hydrogen atom gives a slightly different result because the mass of the particle according to Einstein's mechanics increases with its velocity, approaching the infinite value when the velocity approaches the velocity of light c. In fact, if m_0 is the "rest mass" of a particle (practically, its mass when it moves *much more slowly* than light), the mass at much higher velocity v is given by

$$m = \frac{m_0}{\sqrt{1 - \dfrac{v^2}{c^2}}}$$

which tends to infinity when v approaches c. Since in elliptical motion the velocity varies for the different points of the trajectory (Kepler's Second Law), the mass of the electron varies too, and the calculations become more complicated. In this case, the energies of different orbits corresponding to the same principal quantum number become slightly different and a single level splits into several closely located components. Correspondingly a single spectral line, resulting from the transitions between two quantum levels characterized by two principal quantum numbers m and n, splits into a number of components. This splitting, which can be observed only by using a spectral analyzer with very high dispersive power, is known as the "fine structure" of spectral lines. The frequency differences between the fine structure components depend on the so-called "fine structure constant" α given by

$$\alpha = \frac{e^2}{hc} = \frac{1}{137}$$

This quantity has no physical dimensions, being a pure number, and its smallness accounts for the closeness of fine structure components. If c were infinite, α would be zero and no fine structure at all would be observed.

Another extension of Bohr's original theory came from realization that Sommerfeld's elliptical orbit may not necessarily be in the same plane but may have different orientation in space, making the atoms with many electrons moving along many different orbits look not like flat discs (as in the case of our Solar System) but rather like three-dimensional bodies.

BOHR'S INSTITUTE

Upon Bohr's triumphant return to Denmark, the Royal Danish Academy of Science gave him financial support in order to build his own institute for atomic studies and grant fellowships to young theoretical physicists from all parts of the world who wanted to come to Copenhagen to work with him. Thus, there arose, at the street address Blegdamsvej 15††, a building of the Universitetets Institut for Teoretisk Fysik, and next door to it a director's house for Bohr and his family. It may not be inappropriate to mention here that the Royal Danish Academy of Science draws its main financial support from the Carlsberg Brewery, which produces the best beer in the world. Many years ago the founder of the brewery willed the income from it to the Academy to be used for the development of science, and it was specified in his will that the palatial mansion which the old Carlsberg built for himself right in the middle of his brewery property was to be used as the residence of the most famous living Danish scientist. When Bohr came to his fame and the previous occu-

†† The Institute's official address since has been changed to Blegdamsvej 17.

pant of the Carlsberg Mansion died in the early thirties, Bohr and his family moved into it. In Fig. 14 is given a sketch of a tie which was made for an anniversary of the well-known Danish biochemist Linderstrøm Lang, who for many years was the director of Carlsberg Brewery's research laboratory, and shows a bottle of

Fig. 14. Carlsberg Beer and its consequences.

Carlsberg beer. It is a symbol to everybody who worked on the Carlsberg fellowship in Bohr's Institute. Bohr's Institute quickly became the world center of quantum physics, and to paraphrase the old Romans, "all roads led to Blegdamsvej 17." The Institute buzzed with young theoretical physicists and new ideas about atoms, atomic nuclei, and the quantum theory in general. The popularity of the institute was due both to the genius of its director and his kind, one might say fatherly, heart. Whereas another genius of that era, Albert Einstein, though a very kind man too, never formed what is known as a "school" around him but worked usually with just a single assistant to talk to, Bohr fathered many scientific "children." Almost every country in the world has physicists who proudly say: "I used to work with Bohr." When Bohr once visited the University of Göttingen, he met a young German physicist Werner Heisenberg (see Chapter V), who, at the age of twenty-five, made important advances in the field of quantum mechanics. Bohr proposed that Heisenberg come to Copenhagen to work with him. The next day at the dinner given in Bohr's honor at the University, two uniformed German policemen interrupted the meal and one of them, putting his hand on Bohr's shoulder, announced, "You are arrested on the charge of kidnapping small children!" Of course, the "policemen" were actually two masquerading graduate students, and Bohr never went to jail; but Heisenberg went to Copenhagen! Many theoretical physicists from Europe and America went to Copenhagen for a year, two years, or more and then returned again and again in the later years: P. A. M. Dirac (see Chapter VI) and N. F. Mott (now the director of Cavendish Laboratory) from England; H. A. Kramers and H. Casimir from Holland; Wolfgang Pauli (Chapter III), Werner Heisenberg (Chapter V), and M. Delbrück (see Appendix), and Carl von Weizsäcker from Germany; L. Rosenfeld from Belgium; S. Rosseland from Norway; O. Klein from Sweden; G. Gamow and L. Landau from Russia; R. C. Tolman, J. C. Slater, and J. Robert Oppenheimer from the United States; Y. Nishina from Japan; etc. They went for a long stay, for a short visit, or just for the conferences which were held each spring.

One of the most colorful visitors was Paul Ehrenfest, a professor at the University of Leyden. Ehrenfest was born in Vienna in 1880 and studied under Boltzmann, receiving his Ph.D. in 1904. In that year he married a Russian mathematician, Tatiana, and they moved to St. Petersburg (now known as Leningrad) and stayed there till 1912, when he was invited to the chair of physics at the University of Leyden. He remained there till his death in 1933. His works on statistical mechanics and the theory of adiabatic invariants are too abstract and complicated to be described in this volume, but he was an invaluable member of all scientific meetings because of his broad and deep knowledge of physics and critical turn of mind which helped him to find holes (sometimes the wrong ones) in the newly proposed theory. He liked to call himself a "schoolteacher," and many of his students did very well in their future scientific careers.

Once when I‡‡ was traveling from Denmark to England through Holland, Ehrenfest invited me to stay in his home for a few days. He met me at the station, brought me to his home, and after showing me the guest room in which I was to sleep, said: "No smoking here. If you want to smoke go on the street." At that time I smoked almost as much as today, so I got around

---

‡‡ While relating his own reminiscences the author will depart from the academic convention of modesty and write in the first person.

his regulation by puffing the cigarette smoke into the loading gates of a large Dutch stove in my room. He detested any smell except that of fresh air. One day his student Casimir (now the Scientific Director of Philips Radio Company) had an appointment with him in the afternoon. Before the meeting Cas (a shortening of Casimir, which means "cheese" in Dutch) went to a barbershop to have his hair cut and didn't notice until too late that the barber was rubbing lotion into his blond hair. He had to spend the two hours before his appointment with Ehrenfest walking the streets to let the smell of the lotion dissipate. And, of course, nobody would dare to tell Ehrenfest that Dutch Bols is better (or worse) than English gin!

In an amateur play celebrated among scientists, the Blegdamsvej *Faust* (it is reproduced in English translation in the Appendix to this book), Ehrenfest played the role of Faust whom Mephistopheles (Pauli) seduces by showing him a vision of Gretchen (the neutrino).

The personality of Niels Bohr and the pleasures of life and work in his Institute are still warm in my memories of the years from 1928 to his death, and I hope that a personal anecdote or two will give some of the flavor of that most remarkable man.

After passing the comprehensive examination at the University of Leningrad in the spring of 1928, I managed to get permission from the Soviet Government to spend two months attending summer school at the University of Göttingen. At that time the idea of "proletarian" and "capitalistic" sciences, which are supposed to be hostile to each other, had not developed in Soviet Russia, and the problem of going abroad was just a problem of getting permission to exchange so many Russian rubles into the equivalent amount of German reichsmarks. Presenting recommendations from several University professors, I managed to obtain a rather meager amount of German money and found myself on a ship sailing from Leningrad to German shores. Arrived at Göttingen, I rented a typical student's room and set myself to work.

This was just two years after the discovery of wave mechanics (see Chapter IV), and everyone was busy extending Bohr's original theory of atomic and molecular structure into the new and more advanced field of wave mechanics. But I do not, and never did, like to work in crowded fields, and so decided to see whether something could be done about the structure of the atomic nucleus. At the time the nucleus was being studied experimentally, but no theory of its structure and properties had yet been attempted. During these two months in Göttingen, I struck a gold mine, and was able to explain, on the basis of the wave mechanics, the spontaneous decay of radioactive nuclei as well as nuclear disintegration under the bombardment of particles shot at it from outside. As I found later, a very similar piece of work was done simultaneously by a British physicist, R. W. Gurney, in collaboration with an American physicist, E. U. Condon; in fact, our papers on it were presented for publication almost on the same day.

By the close of summer school in Göttingen my money was coming to an end, and I had to leave for home. But on the way I decided to stop in Copenhagen to meet Professor N. Bohr, whose work I so admired. In Copenhagen I took the cheapest room in a shabby little hotel and went to Bohr's Institute to see his secretary, Frøken (Miss) Schultz, about an appointment. (When I visited Copenhagen several years ago, just about a year before Bohr's death, she was still on the job.) "Professor," said she, "can see you this afternoon." When I went into his study, I found a friendly, smiling, middle-aged man who asked me what my interests in physics were and what I was working on at the moment. So I told him about the work I had done in Göttingen on nuclear transformations, the manuscript of which had been sent in for publication but had not yet appeared. Bohr listened carefully and said: "Very interesting, very, very interesting indeed. How long are you going to stay here?" I explained that I had just enough money left for one more day. "But could you stay for a year," asked Bohr, "if I arrange for you a Carlsberg fellowship of our Academy of Science?"§§ I gasped and finally managed to mumble, "Oh, yes, I could!" Then things moved quickly. Frøken Schultz obtained a very nice room for me in a pension run by Frøken Have, just a few blocks from the Institute, which later became a "campground" for many young physicists coming to work with Bohr. The work in the Institute was very easy and simple: everybody could do whatever he wanted, and come to work and go home whenever he pleased. Another young fellow who came to stay in Frøken Have's pension was Max Delbrück from Germany. We both liked to sleep late in the morning, and Frøken Have devised a special method to get us up. She would come to my room and wake me up, "Dr. Gamow, you'd better get up. Dr. Delbrück has already had his breakfast and left for work!" Then she would carry out the same deception on the still sleeping Delbrück: "Dr. Delbrück, get up. Dr. Gamow has already left for work!" And then Max and I would collide in the bathroom. But still everybody made some progress in work, especially in the evening, which is the most inspiring time for theoretical physicists. This evening work in the Institute's library was often interrupted by Bohr, who would say that he was very tired and would like to go to the movies. The

§§ Of which I have now the honor to be a member.

only movies he liked were wild Westerns (Hollywood style), and he always needed a couple of his students to go with him and explain the complicated plots involving friendly and hostile Indians, brave cowboys, and desperadoes, sheriffs, barmaidens, gold-diggers, and other characters of the Old West. But his theoretical mind showed even in these movie expeditions. He developed a theory to explain why although the villain always draws first, the hero is faster and manages to kill him. This Bohr theory was based on psychology. Since the hero never shoots first, the villain has to decide when to draw, which impedes his action. The hero, on the other hand, acts according to a conditioned reflex and grabs the gun automatically as soon as he sees the villain's hand move. We disagreed with this theory, and the next day I went to a toy store and bought two guns in Western holders. We shot it out with Bohr, he playing the hero, and he "killed" all his students.

Another remark of Bohr, inspired by Western movies, pertained to the theory of probability. "I can believe," he said, "that a girl alone might be walking on a narrow trail somewhere in the Rockies and might lose her step, and, rolling down to the precipice, manage to grab a tiny pine at the brink and so save herself from inevitable death. I can also imagine that, just at that time, a handsome cowboy might be riding the same trail, and, noticing the accident, tie his lasso to his horse's saddle and descend into the precipice to save the girl. But it looks to me extremely improbable that at the same time a cameraman would be present to record this exciting event on film!"

In his youth Niels Bohr was quite an athlete, and on the field of football (played in the Old World with a spherical ball kicked by the foot) was second only to his brother, the well-known mathematician Harald

Bohr, who was the champion hall-keeper of the Copen-
hagen-command.

When during the Christmas vacation in 1930 I went
with Bohr (who was at that time forty-five years old)
to join a group of Norwegian scientists (Rosseland,
Solberg, and "Old Man" Bjerknes) for skiing in the
northern part of Norway beyond the polar circle, Bohr
outskied all of us.

One story I always like to tell or to write on the subject of Bohr is about the evening in Copenhagen when
Bohr, Fru (Mrs.) Bohr, the aforementioned Casimir,
and I were returning from the farewell dinner given
by Oscar Klein on the occasion of his election as a
university professor in his native Sweden. At that late
hour the streets of the city were empty (which cannot
be said of Copenhagen streets today). On the way
home we passed a bank building with walls of large
cement blocks. At the corner of the building the crev-
ices between the courses of blocks were deep enough
to give a toehold to a good alpinist. Casimir, an expert
climber, scrambled up almost to the third floor. When
Cas came down, Bohr, inexperienced as he was, went
up to match the deed. When he was hanging precari-
ously on the second-floor level, and Fru Bohr, Casimir,
and I were anxiously watching his progress, two Copenhagen policemen approached from behind with
their hands on their gun holsters. One of them looked
up and told the other: "Oh, this is only Professor
Bohr!" and they went quietly off to hunt for more dangerous bank robbers.

There is another amusing story illustrating Bohr's
whimsey. Above the front door of his country cottage
in Tisvilde he nailed a horseshoe, which is proverbially
instrumental in bringing luck. Seeing it, a visitor exclaimed: "Being as great a scientist as you are, do you
really believe that a horseshoe above the entrance to

a home brings luck?" "No," answered Bohr, "I certainly do not believe in this superstition. But you know," he added with a smile, "they say that it does bring luck even if you don't believe in it!"

After the discovery of wave mechanics and Heisenberg's formulation of the Uncertainty Principle, Bohr put all his energy into the semi-philosophical development of the *Duality Point of View* on micro-phenomena in physics, according to which every physical entity, be it a light quantum, an electron, or any other atomic particle, presents two sides of a medal. On one side it can be treated as a particle; on the other side as a wave. We will return in Chapter V to a more detailed discussion of this topic. Also working with his assistant, L. Rosenfeld, he extended the original uncertainty relation for a single particle to the case of the electromagnetic field, laying the foundation for a very complicated branch of the Quantum Theory known as quantum electrodynamics.

In the later years, after the discovery of neutrons, Bohr became intensely interested in the then only partially developed theory of nuclear reactions. He showed that when a bombarding particle enters the nuclear interior it does not just kick out some of the nuclear particles as one billiard ball kicks the other, but remains there for a little while (about one ten-billionth of a second perhaps), distributing its impact energy among *all other* particles. Then this energy can be emitted in the form of γ-ray quantum, or, being collected in some of the nuclear particles, kick them out. Thus, for example, the first nuclear reaction studies by Rutherford should not be written as one used to write them:

$$_7N^{14} + {}_2He^4 \longrightarrow {}_8O^{17} + {}_1H^1$$

but rather as a three-step process:

$$_7N^{14} + _2He^4 \longrightarrow _9F^{18[*]} \longrightarrow _8O^{17} + _1H^1$$

In the symbols for the atomic nuclei of various chemical elements the subscripts at the left are the *atomic numbers* of the elements; the superscripts at the right are the *atomic weights* of the isotopes under consideration. The intermediate short-lived product $_9F^{18[*]}$ (the excited nucleus of phoshorus isotope) is known as a "compound nucleus," and the introduction of this notion considerably simplified the analysis of complex nuclear reactions.

When I left Soviet Russia for good in 1933, I became a professor of physics at George Washington University, in Washington, D.C., where, the next year, an old friend and former Bohr student, Dr. Edward Teller, joined me. Following the Copenhagen lead, the Annual Conferences on Theoretical Physics were organized under the auspices of George Washington University and the Carnegie Institution of Washington, where Dr. Merle Tuve was carrying out important experimental studies on nuclear physics. The Conference of 1939 had an especially good attendance with Niels Bohr (who was visiting the United States at that time) and Enrico Fermi (see Chapter VII) sitting in the first row. The first day of the conference passed quietly in discussion of current problems, but the next day brought great excitement. Bohr came a little late that morning, carrying in his hand a radiogram from Dr. Lise Meitner in Stockholm (where she had emigrated from Nazi Germany). The message announced that her former collaborator, Professor Otto Hahn, and his co-workers in Berlin had discovered that barium and another element that turned out to be an isotope of krypton had appeared in a sample of uranium bombarded by neutrons. She and her nephew, the theoretical physicist Otto Frisch, suggested that the experiment showed that a hard-hit uranium nucleus splits into two about equal parts.

The reader can imagine the excitement of that day, and of the remaining days of the conference. That same night the experiment was repeated in Tuve's laboratory, and it was found that the fission of uranium by impact of one single neutron results in the emission of a few more new neutrons. The possibility of a branching chain reaction and the large-scale liberation of nuclear energy seemed open. With the newspaper reporters politely shown from the conference room, the pros and cons of fission chain reaction were carefully weighed. Bohr and Fermi, armed with long pieces of chalk and standing in front of the blackboard, resembled two knights at a medieval tourney. Thus did nuclear energy enter the world of man, leading to uranium fission bombs, nuclear reactors, and later to thermonuclear weapons!

When World War II started, Bohr was in Copenhagen, and he decided to "sit it out" through the Nazi occupation to give as much help as he could to his compatriots. But one day he heard from the Danish underground that he was to be arrested by the Gestapo the next morning. That same night a Danish fisherman rowed him across the Sund to the Swedish shore, where he was picked up by a British Mosquito bomber. Mosquito bombers were small, and the only place for Bohr was the vacant seat at the tail of the plane, usually occupied by a tail gunner. He could communicate with the cockpit only by the intercom. Somewhere over the North Sea the pilot wanted to ask how Bohr felt but couldn't get any response. Thoroughly alarmed, the pilot, on landing at the English airstrip, rushed to the tail of the plane and opened the door of the tail gunner's compartment. There was Bohr, quite safe and sleeping quietly!

Coming from England to the United States, Bohr went directly to Los Alamos to further the work on the fission bomb. Due to the strict security regulations, he carried documents in the name of Nicholas Baker, and was known affectionately as Uncle Nick. There is a story that on one of his visits to Washington he met in the hotel elevator a young woman whom he had often seen in Copenhagen. She used to be the wife of a nuclear physicist, Dr. von Halban, and often visited Copenhagen with her husband. "Very glad to see you again, Professor Bohr," she greeted him. "I am sorry," said Bohr, "you must be mistaken. My name is Nicholas Baker. But," he added, trying to be polite without breaking the security regulations, "I do remember you. You are Mrs. von Halban." "No," she snapped, "I am Mrs. Placzek." The point is that some time previously she had been divorced from her first husband and had married George Placzek, who in earlier years spent a considerable time working with Bohr.

In the summer of 1960, when my wife and I were traveling in Europe, we went to Copenhagen to visit Bohr and his family. He was spending the summer in his country cottage in Tisvilde and invited us to be his guests for a few days. I found him just the same as he was when I saw him first in 1928, but, of course, much slower and less energetic. We had many interesting discussions about the difficulties in the present development of physics. It was therefore quite a shock when, some two years later, I heard on the radio that Niels Bohr had died.

CHAPTER III

W. PAULI AND THE EXCLUSION PRINCIPLE

One of the most colorful visitors at Blegdamsvej was no doubt Wolfgang Pauli. Born in Germany in 1900, he spent most of his life as a professor in Zurich, but would appear unexpectedly as a devil of inspiration wherever theoretical physics was cultivated. His resonating, somewhat sardonic laughter enlivened any conference when he appeared, no matter how dull it was at the start. He always brought along new ideas, telling the audience about them as he continuously walked to and fro along the lecture table, his corpulent body oscillating slightly. His mannerisms inspired someone to write a poem, of which I can recall only this fragment:

> When with colleagues he debates
> All his body oscillates.
> When a thesis he defends
> This vibration never ends.
> Dazzling theories he unveils,
> Bitten from his fingernails!

Once, presumably on a doctor's order, Pauli decided to lose weight and, as in everything else he attempted, succeeded very quickly. When, minus a number of pounds, he appeared again in Copenhagen, he was quite a different man: sad, humorless, and grunting instead of laughing. We all urged him to join us in delicious wiener schnitzel and good Carlsberg beer, and in less than a fortnight Pauli became again his old gay self.

Politically, Pauli was anti-Nazi and would never raise his right hand in the "Heil Hitler" salute—except once. Lecturing at the University of Michigan, in Ann Arbor, he joined a gay boating party on the lake and, stepping out of the boat in the darkness, fell down, breaking his right arm at the shoulder. The arm was put into a cast with a support which held it up at a 45-degree angle. When he appeared next at his lecture he had the chalk in his left hand and was addressing students in the proper Nazi fashion. But he refused to be photographed until the cast was removed.

Pauli started his scientific career very early and, at the age of twenty-one, wrote a book on the Theory of Relativity which (in the revised edition) still represents one of the best books on the subject. He is famous in physics on three counts:

1. The *Pauli Principle,* which he preferred to call The Exclusion Principle.
2. The *Pauli Neutrino,* which he conceived of in

the early twenties and which for three decades escaped experimental detection.

3. The *Pauli Effect,* a mysterious phenomenon which is not, and probably never will, be understood on a purely materialistic basis.

It is well known that theoretical physicists cannot handle experimental equipment; it breaks whenever they touch it. Pauli was such a good theoretical physicist that something usually broke in the lab whenever he merely stepped across the threshold. A mysterious event that did not seem at first to be connected with Pauli's presence once occurred in Professor J. Franck's laboratory in Göttingen. Early one afternoon, without apparent cause, a complicated apparatus for the study of atomic phenomena collapsed. Franck wrote humorously about this to Pauli at his Zurich address and, after some delay, received an answer in an envelope with a Danish stamp. Pauli wrote that he had gone to visit Bohr and at the time of the mishap in Franck's laboratory his train was stopped for a few minutes at the Göttingen railroad station. You may believe this anecdote or not, but there are many other observations concerning the reality of the Pauli Effect!

Quotas for Electron Levels

The Pauli Principle, in contrast to the Pauli Effect, is much better established and pertains to the motion of electrons in atoms. In previous chapters we have described quantum orbits or, in more modern language, quantum vibration states in the Coulomb field of forces surrounding the atomic nucleus.† Since the hydrogen atom contains only one electron, this electron is free to occupy any possible energy state and, in the absence

† See the following chapter.

G. Gamow and W. Pauli on Swiss lake steamer.
(Photographer unknown)

of excitation from outside, sits naturally in the state of lowest energy closest to the nucleus. If it is lifted to some state of higher energy by external force, it drops back to its original lowest state, emitting various lines of the hydrogen spectrum. But what happens in the atoms containing two, three, and more electrons? In Chapter II we derived two formulae for the hydrogen atom in its lowest state ($n = 1$). The radius of the orbit, or, more exactly, the average radius of the continuous function describing this state, is given by

$$r_1 = \frac{h^2}{4\pi^2 e^2 m}$$

and the lowest energy by

$$E_1 = -\frac{4\pi^2 e^4 m}{h^2}$$

These formulae are obtained on the assumption that the electric force is equal to e^2/r^2.

Suppose now that a single electron is revolving around the nucleus with the charge Ze where Z is its atomic number. In this case the force will be Ze^2/r^2 instead of e^2/r^2 and in the foregoing formulae we must substitute Ze^2 for e^2 and Z^2e^4 for e^4. With the increasing atomic number Z the radii of the ground state will decrease as Z, and the absolute values of their energies will increase as Z^2. If, instead of only one electron, we put in the Z electrons, and if they all crowd on the lowest level, the atoms forming the natural system of elements will become smaller and smaller and more and more tightly bound. Of course, carrying out this argument, one must remember that electrostatic repulsion between the electrons will tend to push them apart, but it can easily be shown that this repulsion will not be strong enough to prevent the atoms of heavier elements from shrinking to a considerably smaller size. Thus the atomic volumes‡ would be expected to decrease continuously, and rather rapidly, all the way from hydrogen to uranium as indicated by a broken line in Fig. 15a. The continuous line in the same figure representing experimental data does not look at all like that. It has only a very gentle slope, and is characterized mainly by its saw-shaped form, with the sharp peaks at the positions of the inert gases (He, Ne, A, Kr, Xe, etc.) which, as every chemist knows, are very reluctant to make compounds with other elements or among themselves. Also, if all the electrons of an atom were accumulated on the lowest energy level, the difficulty of extracting one electron from the atom would increase rapidly from the light to the heavy elements along the natural system (broken line in Fig. 15b). This again does not fit at all with the observed curve of the ionization potentials which characterizes that difficulty and is shown by a continuous line in the same figure. And, comparing the two curves, we notice that the maximum difficulty of extracting an atomic electron occurs at the same places where the atomic volumes have the smallest values. Thus, it looks as if the sequence of chemical elements can be visualized as a row of bodies with periodically varying sizes and resistance to the giving away of their electrons. The conclusion results that, with the addition of larger and larger numbers of electrons, the volumes occupied by various quantum states shrink, but that the number of states occupied by the electrons increases so that the total outside diameter of the atom remains approximately constant. Therefore, there must exist some basic physical principle that prevents all the atomic electrons from crowding into the lowest quantum state; as soon

‡ These can be calculated from the known atomic weights and densities of various elements, dividing the weight of 1 cm³ of a given element by the weight of its atoms.

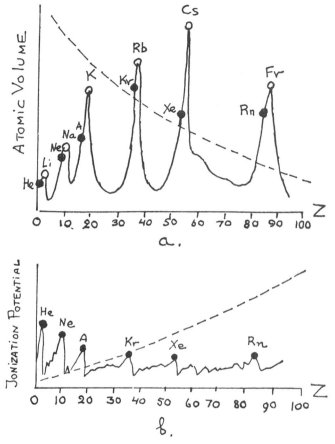

Fig. 15. The changes of atomic volumes and ionization potentials along the natural system of elements. Black circles correspond to noble gases which, being saturated shells, have the highest binding. The black circles are alkali metals which begin to build new shells.

as the "quota" for a given level is filled up, extra electrons must be accommodated on the other quantum states with higher energy. Pauli suggested that things can be settled satisfactorily if one permits *only two electrons* to occupy any given quantum state described by three quantum numbers: radial n_r, azimuthal n_ϕ, and orientational n_o.§ In the original Bohr theory, in the course of which the Pauli Principle was first formulated, these three quantum numbers corresponded to the average diameters, eccentricities, and space orientations of the electron's quantum orbits. In wave mechanics¶ they represent the number of nodes in the complicated three-dimensional oscillatory motion of the ψ functions.

Using the Pauli Principle, Bohr and his co-workers (including, of course, Pauli himself) were able to construct the models of all atoms from hydrogen to uranium. Not only did they explain the periodic changes of atomic volumes and ionization potentials, but also they explained all other properties of atoms, their chemical affinities toward one another, their valencies, and other properties which many years earlier were summarized on a purely empirical basis and systematized by the Russian chemist D. I. Mendeleev, in his periodic system of elements. All these developments fall outside the scope of this little book, the main purpose of which is to describe the revolutionary new ideas rather than their detailed consequences.

The Spinning Electron

The studies and interpretation of atomic spectra on the basis of Bohr's theory, which relied on three quantum numbers (quite natural for three-dimensional space!) to describe the motion of atomic electrons, went happily ahead until, in the early twenties, three

§ We deviate here from conventional notations of quantum numbers for the sake of simplicity. In any branch of science the terminology becomes so cumbersome in the process of its progress that it is very difficult to put it in a simple way for a reader who encounters all these complicated notions for the first time.
¶ See the following chapter.

quantum numbers suddenly turned out to be insufficient. Studies of the Zeeman Effect (the splitting of spectral lines by strong magnetic fields) revealed that there are more components than three integers could account for, and to explain their existence a fourth quantum number was introduced. It was first called the "inner quantum number," a name as good as any because nobody could account for the surplus splitting. Then, in 1925, two Dutch physicists, Samuel Goudsmit and George Uhlenbeck, made a bold proposal. This surplus line-splitting, they suggested, is not due to any additional quantum number describing the electron's orbit in the atom but to the electron itself. Ever since its discovery, the electron was regarded as a point characterized only by its mass and by its electric charge. Why cannot one think of it as a small electrically charged body rotating, like a top, about its axis? It would have a certain angular momentum, and a magnetic moment, as any rotating charge has. A different orientation of the electron's spin (as it was called) in respect to the plane of its orbit would account then for the additional components in line splitting.

It was soon discovered that the proposal works and that, by ascribing to an electron a proper numerical value of spin (that is, angular mechanical momentum) and magnetic moment, one could explain all additional line components found by the experimentalists. The magnetic moment of the spinning electron obligingly came out equal to the so-called *Bohr's magneton*—that is, the minimum amount of magnetism which could be caused by its revolution around the nucleus. But then trouble appeared with the mechanical angular momentum of the spinning electron, which turned out to be *only one-half* of the regular angular momentum $h/2\pi$ of atomic orbits.

Many attempts were made to straighten out this difficulty, and the problem was finally solved by P. A. M. Dirac, in a very unconventional way, only four years later (see Chapter VI). The reason why the introduction of the spinning electron modified the Pauli Principle of atomic electron motion can be understood in the following way. As you will remember, this principle stated that *only two* electrons can occupy any given quantum orbit. Why two? After the discovery of the spinning electron, the original Pauli Principle was amended by the statement: *"only two which possess opposite spin"*; that is, are rotating in opposite directions.

The situation is illustrated graphically in Fig. 16. Drawing (a) represents the old picture in which two point electrons, e_1 and e_2, move along the same orbit. In drawing (b) we have the more recent presentation that two electrons can move along the same orbit only if one of them (e_1) rotates around its axis in the same direction as it revolves around the nucleus, while another (e_2) rotates in the opposite direction. It may be added that drawing (b) is not quite correct since the interaction between the magnetic moment of an electron and the magnetic field within the atom in which it moves changes the orbit slightly so that we actually have two orbits with only one electron on each. (c) Thus the original Pauli Principle can be reformulated by *permitting only one electron on each orbit* if one takes into account the slight splitting of the original orbit.

PAULI AND NUCLEAR PHYSICS

We turn now to an entirely different field of Pauli's activity in science: his contribution to the field of nuclear physics. As everybody knows—or at least should know—radioactive elements emit three kinds of radia-

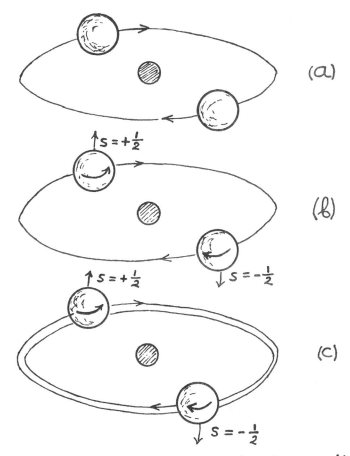

$s = +\frac{1}{2}$

$s = +\frac{1}{2}$ $s = -\frac{1}{2}$

$s = +\frac{1}{2}$

$s = -\frac{1}{2}$

(a)

(b)

(c)

Fig. 16. The motions of two·electrons along the same orbit according to (a) the original Pauli Principle (no more than two electrons can occupy the same orbit); and (b) the amended Pauli Principle (two electrons occupying the same orbit must have opposite spin, i.e., rotate around their axis in the opposite direction); and according to (c) the reformulated Pauli Principle in which, due to magnetic forces arising from the electrons' magnetic momentum, the orbits are not identical and only one electron is permitted on each energy level.

tion: alpha (α), beta (β), and gamma (γ). The principal process of radioactive decay is the emission of α-particles, large chunks of unstable nuclei which were proved by Rutherford to be the nuclei of helium atoms. On the other hand, β-particles are electrons which are sometimes emitted by nuclei following α-decay to restore the balance between the charge and mass upset by ejection of an α-particle. Finally, γ-rays are short electromagnetic waves resulting from the inner disturbances caused by α- and β-emission. For a given radioactive element α-particles have exactly the same energy corresponding to the energy differences of the mother and daughter nuclei. The γ-rays show complex sharp lines, much sharper, in fact, than the lines of optical spectra.

All this activity indicates that atomic nuclei are quantized systems similar to atoms except that they are much smaller; since the nuclei are smaller their transformations, according to the quantum laws, involve much higher energies. But the physicists got a big surprise when James Chadwick discovered in 1914 that, in contrast to α-particles and γ-rays emitted by radioactive nuclei, the β-particles do not have well-defined energies. Quite to the contrary, their energy spectrum extends continuously from practically zero to very large values (Fig. 17). The possibility that this energy spread was due to some internal losses suffered by β-particles in the process of escape from radioactive material was disproved definitely by careful experiments. Thus, one faced a situation in which the nuclear ledgers of income and expenditure of energy did not balance. Niels Bohr, who was first aroused by this experimental finding, took the radical point of view that, if the experiments say so, the Law of Conservation of Energy really does not hold for β-emission or (presumably) for the β-absorption processes. Indeed, this

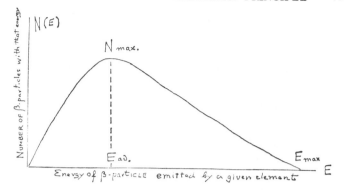

Fig. 17. Typical energy distribution curve of a typical β-emitter.

was the era when so many laws of classical physics were rejected under the impact of the newly developed Theory of Relativity and the Quantum Theory that no law of classical physics seemed to be unshakable. Bohr even tried to use this alleged non-conservation of energy in β-decay processes to explain the seemingly eternal production of energy in stars. According to these little-known and never published views, stars contained in their interior large cores of nuclear matter having the same properties as ordinary atomic nuclei but being much larger (many kilometers rather than 10^{-12} cm in diameter). These stellar cores, which were expected to be unstable, were emitting β-particles of well-defined energy. They were surrounded by ordinary matter in a completely ionized state (plasma, we call it today) consisting of free high-energy electrons and ordinary bare nuclei. The energy of the electrons forming the base of these stellar envelopes was determined by the classical relation

$$E = \frac{3}{2} kT$$

where k is Boltzmann's constant and T the temperature at the base of the envelope.†† On the other hand, the β-particle emitted from the practically flat surface of the nuclear core always had the *same energy* determined by the inner properties of the nuclear fluid. Thus, there must have existed a dynamical equilibrium between the nuclear core and the surrounding ionized gas (plasma) similar to the equilibrium between the water and the saturated vapor above it. The number of β-particles emitted by the radioactive core was equal to the number of free electrons from the envelope absorbed by it. But, whereas the energy of the absorbed free electrons from the envelope was determined by its temperature T, the energy of β-particles emitted by the core was always the same, corresponding to a certain universal nuclear temperature, T_0. Therefore, for $T < T_0$ there was a constant energy flow from the nuclear core into the envelope, and this flow, rising to the star's surface, maintained its high temperature. By virtue of non-conservation of energy in β-emission processes nothing changed in the nuclear core and the stars could shine eternally. Bohr spoke about this theory of his in a slightly critical fashion, but it looked as if he would not be greatly surprised if it were true.

THE NEUTRINO

Pauli, who could not be called conservative in any sense of the word, was nevertheless strongly opposed to Bohr's view. He preferred to assume that the bal-

†† According to the mechanical theory of heat developed by Boltzmann and Maxwell in the middle of the last century, "Heat is nothing but the motion of molecules forming material bodies." They found that the energy of thermal motion (per molecule) is proportional to its absolute temperature—that is, the temperature counted from "absolute zero" at $-273\,°C$. The empirically determined coefficient of proportionality (or rather two-thirds of it) was called the Boltzmann constant.

ance of energy violated by the continuity of β-ray spectra was re-established by the emission of some other kinds of yet unknown particles which he called "neutrons." The name of this "Pauli neutron" was later changed to "neutrino" after Chadwick's discovery of what today we call the neutron. Neutrinos were supposed to be particles carrying no electric charge and having no mass (or, at least, no mass to speak of). They were supposed to be emitted, being paired with β-particles in such a way that the sum of their and the β-particles' energies was always the same, which would of course re-establish the good old law of Conservation of Energy. But, due to their zero-charge and zero-mass, they were practically undetectable, slipping between the fingers of the most skillful experimentalists. Besides Bohr, another neutrinophobe was P. Ehrenfest, and heated verbal discussions and voluminous but never-published correspondence on the subject were exchanged among the three of them.

As the years passed, more and more evidence was accumulated in favor of Pauli's neutrinos even though this evidence was circumstantial. It was not until 1955 that two Los Alamos physicists, F. Reines and C. Cowan, established beyond any doubt the existence of neutrinos by trapping them when they were escaping from the atomic piles of the Savannah River Atomic Energy Commission project. They found that the interaction between neutrinos and matter was so small that an iron shield several light years thick would be needed to reduce the intensity of the neutrino's beam by one-half. Today neutrinos have a larger and larger place in the study of elementary particles and astrophysical phenomena; they may become the most important elementary particles in physics. Like electrons, neutrinos were found to behave as little spinning tops, and their angular momenta are exactly equal to an

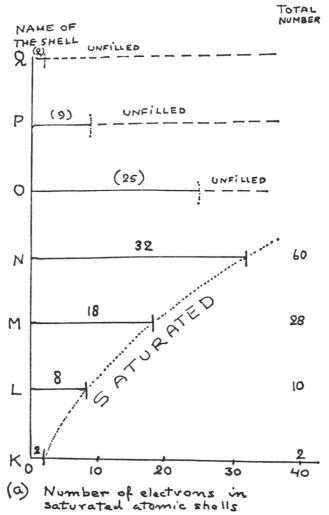

(a) Number of electrons in saturated atomic shells

Fig. 18. A comparison between (a) the saturation of electron shells in Bohr-Coster diagram of the sequence of atoms and (b) Mayer-Jensen's diagram of the saturation of proton and neutron shells in the sequence of atomic nuclei.

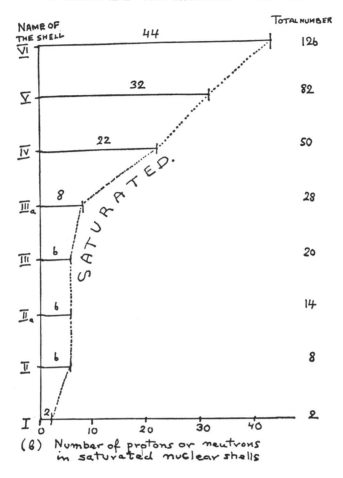

(8) Number of protons or neutrons in saturated nuclear shells

electron's. But since neutrinos carry no electric charge, their magnetic moment is equal to zero.

It was later found experimentally that protons and neutrons also have the same spin as electrons and also obey the Pauli Principle. The latter fact is of great importance in the problem of internal structure of atomic nuclei, which are formed by an agglomeration of various numbers of protons and neutrons tightly bound together by nuclear forces. As was first shown by G.

Gamow in 1934, the natural sequence of atomic nuclei from hydrogen to uranium isotopes shows periodic changes in their various properties, similar to but much smaller than the changes of chemical properties of atoms in Mendeleev's periodic system of elements. This periodicity indicated that atomic nuclei must have a shell structure similar to but probably more complicated than the shell structure of the atomic electron envelopes. The situation here is complicated by the fact that, whereas atomic envelopes are formed by only one kind of particle, namely electrons, the nuclei are formed by two kinds of particles, neutrons and protons, and that Pauli's Exclusion Principle applies to each kind separately. Thus, any given energy state characterized by three quantum numbers can accommodate two protons (with opposite spin) along with two neutrons (also with opposite spin), and we actually have two systems of shells, one for protons and one for neutrons, overlapping on one another. There is another difficulty. Because of the close packing of protons and neutrons in the nucleus, the calculations of energy levels become considerably more complicated. The problem was finally solved in 1949 by M. Goeppert Mayer and H. Jensen et al., who were able to prove that the neutron as well as the proton shells inside the nuclei have capacities of 2, 8, 14, 20, 28, 50, 82, and 126 particles each, as is shown schematically in Fig. 18. These numbers, known as "Magic Numbers," permitted physicists to understand completely the observed periodicity in nuclear structure.

Another important application of Pauli's Principle can be found in the work of P. A. M. Dirac, who used it for the explanation of the stability of matter, as will be described in Chapter VI. On the basis of his theory, Dirac was led to the conclusion that to each "normal particle," such as an electron, proton, neutron, and the hordes of other particles discovered during the last decade, there must exist an "anti-particle" with exactly the same physical properties but the opposite electric charge. This will be discussed in more detail in Chapters VI and VIII.

To finish this present chapter, it is enough to say that it is just as difficult to find the branch of modern physics in which the Pauli Principle is not used as to find a man as gifted, amiable, and amusing as Wolfgang Pauli was.

CHAPTER IV

L. DE BROGLIE AND PILOT WAVES

Louis Victor, Duc de Broglie, born in Dieppe in 1892, who became the Prince de Broglie on the death of an elder brother, had a rather unusual scientific career. As a student at the Sorbonne he decided to devote his life to medieval history, but the onset of World War I induced him to enlist in the French Army. Being an educated man, he got a position in one of the field radio-communication units, a novelty at that time, and turned his interest from Gothic cathedrals to electro-magnetic waves. In 1925 he presented a doctoral thesis which contained such revolutionary ideas concerning a modification of the Bohr original theory of atomic structure that most physicists were rather skeptical; some wit, in fact, dubbed de Broglie's theory *"la Comédie Française."*

Having worked with radio waves during the war, and being a connoisseur of chamber music, de Broglie chose to look at an atom as some kind of musical instrument which, depending on the way it is constructed, can emit a certain basic tone and a sequence of overtones. Since by that time Bohr's electronic orbits were fairly well established as characterizing different quantum states of an atom, he chose them as a basic pattern for his wave scheme. He imagined that each electron moving along a given orbit is accompanied by some mysterious pilot waves (now known as de Broglie waves) spreading out all along the orbit. The first quantum orbit carried only one wave, the second two waves, the third three, etc. Thus the length of the first wave must be equal to the length $2\pi r_1$ of the first quantum orbit, the length of the second wave must be equal to one-half of the length of the second orbit, $\frac{1}{2} \cdot 2\pi r_2$, etc. In general, the nth quantum orbit carries n waves with the length $\frac{1}{n} 2\pi r_n$ each.

As we have seen in Chapter II, the radius of the nth orbit in Bohr's atom is

$$r_n = \frac{1}{4\pi^2} \frac{h^2}{me^2} n^2$$

From the equality of the centrifugal force due to the orbital motion, and the electrostatic attraction beween the charged particles, we obtain:

$$\frac{mv_n^2}{r_n} = \frac{e^2}{r_n^2}$$

or

$$e^2 = mv_n^2 r_n$$

Substituting this value of e^2 into the original formula, we get

$$r_n = \frac{1}{4\pi^2} \frac{h^2 n^2}{m} \cdot \frac{1}{m v_n^2 r_n}$$

or

$$(2\pi r_n)^2 = \frac{h^2 n^2}{m^2 v^2}$$

Extracting the square root from both sides of this equation we finally obtain:

$$2\pi r_n = n \cdot \frac{h}{m v_n}$$

Thus, if the length λ of the wave accompanying an electron is equal to Planck's constant h divided by the mechanical momentum mv of the particle, then

$$\lambda = \frac{h}{mv}$$

and de Broglie could satisfy his desire to introduce waves of such a nature that *1, 2, 3, etc., of them would fit exactly into the 1st, 2nd, 3rd of Bohr's quantum orbits* (Fig. 19). The result given is mathematically equivalent to Bohr's original quantum condition and brings in nothing physically new—nothing, that is, but *an idea:* the motion of the electrons along Bohr's quantum orbits is accompanied by mysterious waves of the lengths determined by the mass and the velocity of the moving particles. If these waves represented some kind of physical reality, they should also accompany particles moving freely through space, in which case their existence or non-existence could be checked by direct experiment. In fact, if the motion of electrons is always guided by de Broglie waves, a beam of electrons under proper conditions should show diffraction phenomena similar to those characteristic of beams of light.

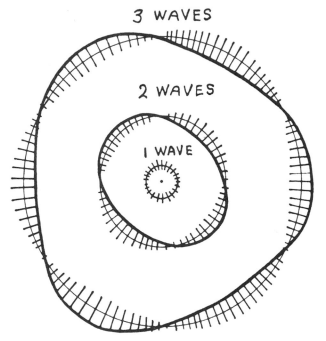

Fig. 19. De Broglie waves fitted to quantum orbits in Bohr's atom model.

Electron beams accelerated by electron tensions of several kilovolts (which are commonly used in laboratory experiments) should, according to de Broglie's formula, be accompanied by pilot waves of about 10^{-8} cm wavelength, which is comparable to the wavelength of ordinary X-rays. This wavelength is too short to show a diffraction in ordinary optical gratings and should be studied with the technique of standard X-ray spectroscopy. In this method the incident beam is reflected from the surface of a crystal, and the neighboring crystalline layers, located about 10^{-8} cm apart, have the function of the more widely separated lines in optical diffraction gratings (Fig. 20). This experiment was carried out simultaneously and independently by

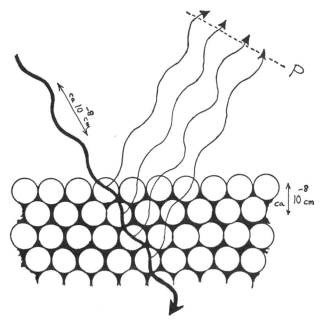

Fig. 20. An incident wave, be it a short electromagnetic wave (X-ray) or a de Broglie wave associated with a beam of fast electrons, produces wavelets as it passes through the successive layers of a crystal lattice. Depending on the angle of incidence, dark and light interference fringes appear. (P is the phase plane.)

Sir George Thomson (son of Sir J. J. Thomson) in England, and G. Davisson and L. H. Germer in the United States, who used a crystal arrangement similar to that of Bragg and Bragg, but substituted for the beam of X-rays a beam of electrons moving at a given velocity. In the experiments a characteristic diffraction pattern appeared on the screen (or photographic plate) that was placed in the way of the reflected beam, and the diffraction bands widened or narrowed when the velocity of incident electrons was increased or decreased. The measured wavelength coincided exactly in all cases with that given by the de Broglie formula. Thus the de Broglie waves became an indisputable physical reality, although nobody understood what they were.

Later on a German physicist, Otto Stern, proved the existence of the diffraction phenomena in the case of atomic beams. Since atoms are thousands of times more massive than electrons, their de Broglie waves were expected to be correspondingly shorter for the same velocity. To make atomic de Broglie waves of a length comparable with the distances between the crystalline layers (about 10^{-8} cm), Stern decided to use the thermal motion of atoms, since he could regulate the velocity simply by changing the temperature of the gas. The source consisted of a ceramic cylinder heated by an electric wire wound around it. At one end of the closed cylinder was a tiny hole through which the atoms escaped at their thermal velocity into a much larger evacuated container, and in their flight through space they hit a crystal placed in their way. The atoms reflected in different directions stuck to metal plates cooled by liquid air, and the number of atoms on the different plates was counted by a complicated method of chemical microanalysis. Plotting the number of atoms scattered in different directions against the scattering angle, Stern obtained again a perfect diffraction pattern corresponding exactly to the wavelength calculated from de Broglie's formula. And the bands became wider or thinner when the temperature of the cylinder was changed.

When in the late twenties I was working at Cambridge University with Rutherford, I decided to spend Christmas vacation in Paris (where I had never been before) and wrote to de Broglie, saying that I would like very much to meet him and to discuss some problems of the Quantum Theory. He answered that the

University would be closed but that he would be glad to see me in his home. He lived in a magnificent family mansion in the fashionable Parisian suburb Neuilly-sur-Seine. The door was opened by an impressive-looking butler.

"*Je veux voir Professeur de Broglie.*"

"*Vous voulez dire, Monsieur le Duc de Broglie,*" retorted the butler.

"O.K., *le Duc de Broglie,*" said I, and was let into the house.

De Broglie, wearing a silk house coat, met me in his sumptuously furnished study, and we started talking physics. He did not speak any English; my French was rather poor. But somehow, partly by using my broken French and partly by writing formulas on paper, I managed to convey to him what I wanted to say and to understand his comments. Less than a year later, de Broglie came to London to deliver a lecture at the Royal Society, and I was, of course, in the audience. He delivered a brilliant lecture, in perfect English, with only a slight French accent. Then I understood another of his principles: when foreigners come to France they must speak French.

A number of years later when I was planning a trip to Europe and de Broglie asked me to deliver a special lecture in the institute of Henri Poincaré, of which he was a director, I decided to come well prepared. I planned to write the lecture down in my (still) poor French on board the liner crossing the Atlantic, have somebody in Paris correct the text, and use it as notes at the lecture. But, as everybody knows, all good resolutions collapse on an ocean voyage offering many distractions, and I had to face the audience in the Sorbonne completely unprepared. The lecture went through somewhat stumblingly, but my French held, and everybody understood what I had to say. After the lecture

I told de Broglie that I was sorry that I did not carry out my original plan of having the corrected French notes. *"Mon Dieu!"* he exclaimed, "it is lucky that you didn't."

De Broglie told me about a lecture delivered by the noted British physicist R. H. Fowler. It is well known that since English is the best language in the world, the English are of the opinion that all foreigners should learn it, thus freeing themselves from the need to learn anyone else's language. Since the lecture in the Sorbonne had to be in French, Fowler had prepared the complete English text of his lecture, and he sent it well in advance to de Broglie, who had personally translated it into French. Thus Fowler lectured in French, using the typewritten French text. De Broglie said that after the lecture a group of students came to him, *"Monsieur le Professeur,"* they said, "we are greatly puzzled. We expected that Professor Fowler would lecture in English, and we all know enough English to be able to understand. But he did not speak English but some other language and we cannot figure out what language it was." *"Et parfois!"* added de Broglie, "I had to tell them that Professor Fowler was lecturing in French!"

SCHRÖDINGER'S WAVE EQUATION

Creating the revolutionary idea that the motion of atomic particles is guided by some mysterious pilot waves, de Broglie was too slow to develop a strict mathematical theory of this phenomenon, and, in 1926, about a year after his publication, there appeared an article by an Austrian physicist, Erwin Schrödinger, who wrote a general equation for de Broglie waves and proved its validity for all kinds of electron motion. While de Broglie's model of the atom resembled more an unusual stringed instrument, or rather a set of vibrating concentrical metal rings of different diameters, Schrödinger's model was a closer analogy to wind instruments; in his atom the vibrations occur throughout the entire space surrounding the atomic nucleus.

Consider a flat metal disc something like a cymbal fastened in the center (Fig. 21a). If one strikes it, it

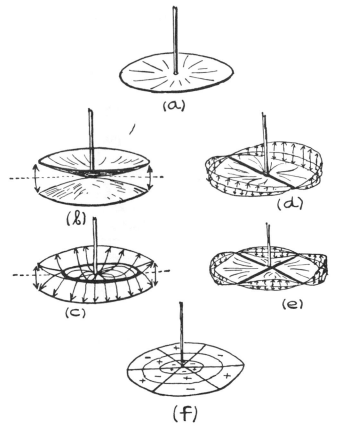

Fig. 21. Various vibration modes of an elastic disc fastened in the center: (a) state of rest; (b) nodal point in the center; (c) one circular nodal line; (d) one radial nodal line; (e) two radial nodal lines; (f) three radial and two circular nodal lines.

will begin to vibrate with its rim moving periodically up and down (Fig. 21b). There exist also more complicated kinds of vibrations (overtones) like the pattern shown in Fig. 21c where the center of the plate and all the points located along the circle between the center and circumference (marked by heavy line in the figure) are at rest, so that, when the material bulges up within that circle the material outside the circle moves down, and vice versa. The motionless points and lines of a vibrating elastic surface are called the *nodal* points and lines; one can extend Fig. 21c by drawing higher overtones which correspond to two or more nodal circles around the central nodal point.

Besides such "radial" vibrations there also exist the "azimuthal vibrations" in which the nodal lines are straight lines passing through the center as shown in Fig. 21d and e, where arrows indicate whether the membrane lifts or sinks in respect to the equilibrium horizontal position. Of course, the radial and azimuthal vibrations can exist simultaneously in a given membrane. The resulting complex state of motion should be described by two integers n_r and n_ϕ, giving the numbers of radial and azimuthal nodal lines.

Next in complexity are the three-dimensional vibrations such as, for example, the sound waves in the air filling a rigid metal sphere. In this case it becomes necessary to introduce the third kind of nodal lines and also the third integer n_l giving their number.

This kind of vibrations was studied in theoretical acoustics many years ago, and, in particular, Hermann von Helmholtz in the last century made detailed studies of the vibrations of air enclosed in rigid metal spheres (Helmholtz resonators). He drilled a little hole in the sphere, to let in sound from the outside, and used a siren which emitted a pure tone, the pitch of which could be changed continuously by changing the rotation speed of the siren's disc. When the frequency of the siren's sound coincided with one of the possible vibrations of air in the sphere, resonance was observed. These experiments stood in perfect agreement with the mathematical solutions of the wave equation for sound, which is too complicated to be discussed in this book.

The equation written by Schrödinger for de Broglie's waves is very similar to the well-known wave equations for the propagation of sound and light (that is, electromagnetic) waves, except that for a few years there remained the mystery of just *what was vibrating*. We will return to this question in the next chapter.

When an electron moves around a proton in a hydrogen atom the situation is somewhat similar to the vibration of gas within a rigid spherical enclosure. But whereas for Helmholtz vibrators there is a rigid wall preventing the gas from expanding beyond it, the atomic electron is subject to electric attraction of the central nucleus which slows down the motion when the electron travels farther and farther from the center, and stops it when it goes beyond the limit permitted by its kinetic energy. The situation in both cases is shown graphically in Fig. 22. In the figure on the left the "potential hole" (that is, the lowering of potential energy in the neighborhood of a certain point) resembles a cylindrical well; the figure on the right looks more like a funnel-shaped hole in the ground. The horizontal lines represent the quantized energy levels, the lowest of them corresponding to the lowest energy the particle can have. Comparing Fig. 22b with Fig. 12 of Chapter II, we find that the levels of the hydrogen atom calculated on the basis of Schrödinger's equation are identical with those obtained from Bohr's old theory of quantum orbits. But the physical aspect is quite different. Instead of sharp circular and elliptic

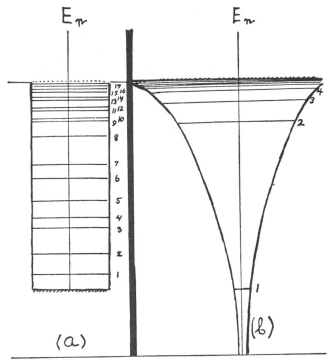

E_n E_n

(a) (b)

Fig. 22. *Quantum energy levels in a rectangular potential hole (a) and in a funnel-shaped potential hole (b).*

orbits along which point-shaped electrons run, we have now a full-bodied atom represented by multishaped vibrations of something which in the early years of wave mechanics was called, for lack of a better name, a ψ-function (Greek letter *psi*).

It must be remarked here that the rectangular well potential distribution shown in Fig. 22a turned out to be very useful for a description of proton and neutron motion within the atomic nucleus, and later was used successfully by Maria Goeppert Mayer and, independently, by Hans Jensen for an understanding of the energy levels within the atomic nuclei and the origin of γ-ray spectra of radioactive nuclear species.

The frequencies of different ψ-vibration modes do not correspond to the frequencies of the light wave emitted by the atom, but to the energy values of the different quantum states divided by h. Thus, the emission of a spectral line necessitated the excitation of two vibration modes, say ψ_m and ψ_n, with the resulting composite frequency

$$v_{m,n} = \frac{E_m}{h} - \frac{E_n}{h} = \frac{E_m - E_n}{h}$$

which is the same as Bohr's expression for the frequency of light quantum resulting from the transition of the atomic electron from the energy level E_m to the lower energy level E_n.

Applying Wave Mechanics

Apart from giving a more rational foundation to Bohr's original idea of quantum orbits, and removing some discrepancies, wave mechanics could explain some phenomena well beyond the reach of the old Quantum Theory. As was mentioned in Chapter II, the author of the present book and, independently, a team consisting of Ronald Gurney and Edward Condon successfully applied Schrödinger's wave equation to the explanation of the emission of α-particles by radioactive elements, and their penetration into the nuclei of other lighter elements with the resulting transformation of elements. To understand this rather complicated phenomenon, we will compare an atomic nucleus to a fortress surrounded on all sides by high walls; in nuclear physics the analogy of the fortress walls is known as a *potential barrier*. Due to the fact that both the atomic nucleus and the α-particle carry a positive electric charge, there exists a strong repulsive Coulomb force† acting on the α-particle approaching a nucleus. Under the action of that force an α-particle shot at the nucleus may be stopped and thrown back before it comes into direct contact with the nucleus. On the other hand, α-particles that are inside the various nuclei as constituent parts of them are prevented from escaping by very strong attractive nuclear forces (analogous to the cohesion forces in ordinary liquids), but these nuclear forces act only when the particles are closely packed, being in direct contact with one another. The combination of these two forces forms a potential barrier preventing the inside particles from getting out and the outside particles from getting in, unless their kinetic energy is high enough to climb over the top of the potential barrier.

Rutherford found experimentally that the α-particles emitted by various radioactive elements, such as uranium and radium, have much smaller kinetic energy than that needed to get out over the top of the barrier. It was also known that when α-particles are shot at the nuclei from outside with less kinetic energy than needed to reach the top of the potential barrier they often penetrate into the nuclei, producing artificial nuclear transformations. According to the basic principles of classical mechanics, both phenomena were absolutely impossible, so that no spontaneous nuclear decay resulting in the emission of α-particles, and no artificial nuclear transformations under the influence of α-bombardment could possibly exist. And yet both were experimentally observed!

If one looks on the situation from the point of view

† During the early studies of electric phenomena, the French physicist Charles de Coulomb found that forces acting between charged particles are proportional to the product of their electric charges and inversely proportional to the square of the distance between them. This is known as Coulomb's Law.

of wave mechanics, it appears quite different, since the motion of the particles is governed by de Broglie's pilot waves. To understand how wave mechanics explains these classically impossible events, one should remember that wave mechanics stands in the same relation to the classical Newtonian mechanics as wave optics to the old geometrical optics. According to Snell's Law, a light ray falling on a glass surface at a certain incidence angle i (Fig. 23a) is refracted at a smaller angle r, satisfying the condition $sin\ i/sin\ r = n$ where n is the refractive index of glass. If we reverse the situation (Fig. 23b), and let a light ray propagating through glass exit into the air, the angle of refraction will be larger than that of incidence and we will have $sin\ i/sin\ r = 1/n$. Thus a light ray falling on the interface between the glass and air at an angle of incidence greater than a certain critical value will not enter into the air at all but will be totally reflected back into the glass. According to the wave theory of light the situation is different. Light waves undergoing total internal reflection are not reflected from the mathematical boundary between the two substances, but penetrate into the second medium (in this case air) to the depth of several wavelengths λ and then are thrown back into the original medium (Fig. 23c). Therefore, if we place another plate of glass a few wavelengths away (a few microns, in the case of visible light), some amount of light coming into the air will reach the surface of that glass and continue to propagate in the original direction (Fig. 23d). The theory of this phenomenon can be found in the books on optics published a century ago and represents a standard demonstration in many university courses on optics.

Similarly, de Broglie waves which guide the motion of α-particles and other atomic projectiles can pene-

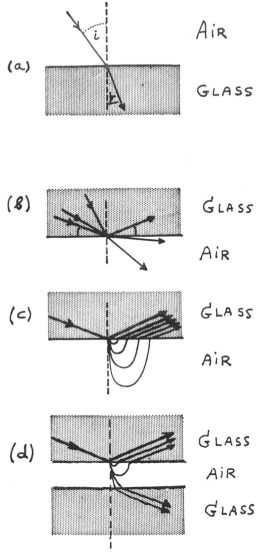

Fig. 23. *Analogy Between Wave Mechanics and Wave Optics. In (a) we have the familiar picture of refraction of light entering from the rarer into the denser medium. In (b) we have the reverse case when the light entering from the denser into the rarer medium can be completely reflected from the interface if the angle of incidence exceeds*

trate through the regions of space which are prohibited to particles by classical Newtonian mechanics, and α-particles, protons, etc., can cross the potential barriers whose height is greater than the energy of the incident particle. But the probability of penetration is of physical importance only for particles of atomic mass, and for barriers not more than 10^{-12} or 10^{-13} cm wide. Let us take, for example, a uranium nucleus which emits an α-particle after an interval of about 10^{10} years. An α-particle imprisoned within the uranium potential barrier hits the barrier wall some 10^{21} times per second, which means that the chance of escape after a simple hit is one out of $10^{10} \times 3 \cdot 10^7 \times 10^{21} \cong 3 \cdot 10^{38}$ hits (here $3 \cdot 10^7$ is the number of seconds in a year). Similarly, the chances that an atomic projectile will enter the nucleus are very small for each individual hit, but may become considerable if a very large number of nuclear collisions are involved. It was shown in 1929 by Fritz Houtermans and Robert Atkinson that the nuclear collisions caused by intensive thermal motion, known as *thermonuclear reactions*, are responsible for the production of energy in the Sun and stars. Physicists are working hard now to produce the so-called "con-

---

a certain critical value. According to the wave theory of light, the reflection takes place not on the mathematical surface separating the two media, but within a certain layer several wavelengths thick. Thus, if the second layer of the denser medium is placed a few wavelengths beyond the first layer, a fraction of the incident light will not be totally reflected but will penetrate into the second dense layer propagating in the original direction. Similarly, according to wave mechanics, some particles can penetrate through the regions prohibited by classical mechanics, where the potential is higher than the original kinetic energy of the particles.

trolled thermonuclear reactions" which would supply us with cheap, inexhaustible, and harmless sources of nuclear energy. All this would have been impossible if Newton's classical mechanics had not been replaced by de Broglie-Schrödinger wave mechanics.

CHAPTER V

W. HEISENBERG AND THE
UNCERTAINTY PRINCIPLE

Simultaneously with Schrödinger's paper on wave mechanics, which appeared in the *Annalen der Physik,* there appeared in another German magazine, the *Physikalische Zeitschrift,* a paper by Werner Heisenberg of Göttingen University dedicated to the same subject and leading to exactly the same results. But, to the great astonishment of the physicists who read these two papers, they started from entirely different physical assumptions, used entirely different mathematical methods, and seemed to have nothing to do with each other.

As was described in the previous chapter, Schrödinger visualized the motion of atomic electrons as being governed by a system of generalized three-dimensional de Broglie waves surrounding the atomic nucleus, whose shapes and vibration frequencies were determined by the field of electric and magnetic forces. Heisenberg, on the other hand, devised a more abstract model. He treated the atom as if it were composed of an infinite number of linear "virtual" vibrators with frequencies coinciding with all possible frequencies that the atom in question could emit. Thus, whereas in Schrödinger's picture the emission of a spectral line with the frequency $\nu_{m,n}$ was considered a "cooperative result" of two vibration functions ψ_m and ψ_n,† in Heisenberg's model the same spectral line was emitted by an individual vibrator which we may call $V_{m,n}$.

In classical mechanics a linear vibrator is described by two numbers: its displacement from the position of equilibrium q and its velocity v, both quantities changing periodically in time. It is customary in advanced mechanics to use, instead of the velocity v, the mechanical momentum‡ p, defined as the product of the particle's mass by its velocity $(p = mv)$. For a given law of force acting on the vibrator, it will possess a well-defined frequency ν. But the optical spectrum possesses

† For simplicity's sake we use here only one instead of three quantum numbers for each vibration mode.

‡ The notion of *mechanical momentum,* which Isaac Newton called *amount of motion,* was introduced in his book *Mathematical Principles of Natural Philosophy* and resulted from combination of the Second and Third Laws of Motion. If two particles, originally at rest, interact with one another, the acting forces F_1 and F_2 are numerically equal and oppositely directed. On the other hand, the velocities gained during the period of interaction (v_1 and v_2) are inversely proportional to the masses (m_1 and m_2) of the two particles. Thus the "amounts of motion" (or mechanical momenta as we call it today) are numerically equal and oppositely directed. This is the famous *Law of the Conservation of Mechanical Momentum.*

a double manifold of frequencies that can be represented by a table:

$\nu_{m,n}$	ν_{11}	ν_{12}	ν_{13}	ν_{14}	ν_{15}	ν_{16}	etc.
	ν_{21}	ν_{22}	ν_{23}	ν_{24}	ν_{25}	ν_{26}	etc.
	ν_{31}	ν_{32}	ν_{33}	ν_{34}	ν_{35}	ν_{36}	etc.
	ν_{41}	ν_{42}	ν_{43}	ν_{44}	ν_{45}	ν_{46}	etc.
	etc.	etc.	etc.	etc.	etc.	etc.	etc.

Number arrays of this kind were known for a long time to mathematicians as *matrices* and were used successfully in the solution of various algebraic problems. Matrices can be *finite* if the indices m and n run from 1 to a given number, or *infinite* if both m and n run to infinity. In fact, there was developed a special branch of mathematics in which any given matrix (finite or infinite) can be represented by just one symbol printed in bold type. Thus **a** meant a matrix:

a_{11}	a_{12}	a_{13}	a_{14}	a_{15}	a_{16}	a_{17}	etc.
a_{21}	a_{22}	a_{23}	a_{24}	a_{25}	a_{26}	a_{27}	etc.
a_{31}	a_{32}	a_{33}	a_{34}	a_{35}	a_{36}	a_{37}	etc.
etc.	etc.	etc.	etc.	etc.	etc.	etc.	etc.

Like ordinary numbers, matrices can be added, subtracted, multiplied, and divided by one another. The rules of addition and subtraction are similar to those of ordinary numbers: one adds or subtracts, term by term. For example:

$$\mathbf{a} \pm \mathbf{b} =$$

$a_{11} \pm b_{11}$	$a_{12} \pm b_{12}$	$a_{13} \pm b_{13}$	etc.
$a_{21} \pm b_{21}$	$a_{22} \pm b_{22}$	$a_{23} \pm b_{23}$	etc.
$a_{31} \pm b_{31}$	$a_{32} \pm b_{32}$	$a_{33} \pm b_{33}$	etc.
etc.	etc.	etc.	etc.

From this definition it follows that the addition of matrices is subject to a commutative law; that is, $\mathbf{a} + \mathbf{b} = \mathbf{b} + \mathbf{a}$, just as $3 + 7 = 7 + 3$ or $a + b = b + a$. But the laws of multiplication and division of matrices are more complicated. To obtain the term in the mth line and nth column of the product $\mathbf{ab}$, one has to multiply term by term the entire sequence of terms in the mth line of $\mathbf{a}$ by the entire sequence of terms in the nth column of $\mathbf{b}$ and add all these products together. Schematically this rule can be represented by the following diagrams where a circled dot of the product is given by the sum of products of dots placed in the squares:

or:

To become better acquainted with this procedure let us use numbers rather than letters for the elements of the matrix, and calculate the product of two matrices:

$$
\text{For }
\begin{vmatrix}
1 & 3 & 5 \\
2 & 5 & 1 \\
4 & 3 & 2
\end{vmatrix}
\times
\begin{vmatrix}
3 & 5 & 4 \\
1 & 1 & 1 \\
2 & 3 & 5
\end{vmatrix}
\text{ we get: }
\begin{vmatrix}
16 & 23 & 32 \\
13 & 18 & 18 \\
19 & 29 & 29
\end{vmatrix}
$$

because $1 \times 3 + 3 \times 1 + 5 \times 2 = 16$; $1 \times 5 + 3 \times 1 + 5 \times 3 = 23$; etc.

Let us now reverse the order of multiplicands, and calculate:

$$
\text{For }
\begin{vmatrix}
3 & 5 & 4 \\
1 & 1 & 1 \\
2 & 3 & 5
\end{vmatrix}
\times
\begin{vmatrix}
1 & 3 & 5 \\
2 & 5 & 1 \\
4 & 3 & 2
\end{vmatrix}
\text{ we get: }
\begin{vmatrix}
29 & 46 & 28 \\
7 & 11 & 8 \\
28 & 36 & 23
\end{vmatrix}
$$

The result is quite different from the first case! *The commutative law of multiplication, so common in arithmetic and ordinary algebra, does not hold in matrix calculus!* That is why calculation with matrices is called *non-commutative algebra.* It must be remarked here that not all pairs of matrices necessarily give different results when the order of multiplication is reversed. If the result is the same, one says that these two matrices *commute;* if the result is different they *do not commute.*

The division of the matrices is defined in the same way as in ordinary algebra where $\frac{a}{b} = a \cdot \frac{1}{b}$ and the value of $\frac{1}{b}$ (the reverse of that number) is subject to the condition $b \cdot \frac{1}{b} = 1$. In non-commutative algebra $\frac{a}{b} = a \cdot \frac{1}{b}$ where $\frac{1}{b}$ satisfies the condition $b \cdot \frac{1}{b} = 1$ and

$$\mathbf{1} = \begin{array}{cccccc} 1 & 0 & 0 & 0 & 0 & \text{etc.} \\ 0 & 1 & 0 & 0 & 0 & \text{etc.} \\ 0 & 0 & 1 & 0 & 0 & \text{etc.} \\ \text{etc.} & \text{etc.} & \text{etc.} & \text{etc.} & \text{etc.} & \text{etc.} \end{array}$$

Heisenberg's idea was that, since the frequencies of the spectral lines emitted by an atom represent an infinite matrix:

$$\nu_{m,n} \begin{array}{ccccc} \nu_{11} & \nu_{12} & \nu_{13} & \nu_{14} & \text{etc.} \\ \nu_{21} & \nu_{22} & \nu_{23} & \nu_{24} & \text{etc.} \\ \nu_{31} & \nu_{32} & \nu_{33} & \nu_{34} & \text{etc.} \\ \text{etc.} & \text{etc.} & \text{etc.} & \text{etc.} & \text{etc.} \end{array}$$

The mechanical quantities, such as velocities, momenta, etc., should also be written in the form of matrices. Thus, the mechanical momenta and coordinates should be given by the matrices:

$$\mathbf{p} = \begin{array}{lllll} p_{11} & p_{12} & p_{13} & p_{14} & \text{etc.} \\ p_{21} & p_{22} & p_{23} & p_{24} & \text{etc.} \\ p_{31} & p_{32} & p_{33} & p_{34} & \text{etc.} \\ \text{etc.} & \text{etc.} & \text{etc.} & \text{etc.} & \text{etc.} \end{array}$$

and

$$\mathbf{q} \begin{array}{lllll} q_{11} & q_{12} & q_{13} & q_{14} & \text{etc.} \\ q_{21} & q_{22} & q_{23} & q_{24} & \text{etc.} \\ q_{31} & q_{32} & q_{33} & q_{34} & \text{etc.} \\ \text{etc.} & \text{etc.} & \text{etc.} & \text{etc.} & \text{etc.} \end{array}$$

where the individual values of p_{mn}'s and q_{mn}'s oscillate with the frequencies ν_{mn} given in the table above.

Substituting p and q into the equations of classical mechanics, Heisenberg expected to obtain individual frequencies and amplitudes of various "virtual" vibrators. There was, however, one more step necessary to obtain the final result. In classical mechanics p and q are ordinary numbers, and therefore it makes no difference in calculations whether one writes pq or qp. The matrices $\mathbf{p}$ and $\mathbf{q}$ do not commute ($\mathbf{pq} \neq \mathbf{qp}$), so that one must introduce an additional postulate stating what the difference between $\mathbf{pq}$ and $\mathbf{qp}$ is. Heisenberg assumed that this difference, which is also a matrix, is equal to the unit matrix I with an ordinary numerical coefficient for which he chose $h/2\pi i$. Thus, the additional condition became:

$$pq - qp = \frac{h}{2\pi i} I$$

Adding this condition to the classical equation of mechanics written in the matrix form, he obtained a system of equations which led to the correct values of the frequencies and relative intensities of spectral lines, which were identical with the results obtained by Schrödinger through using his wave equation.

The unexpected identity of the results obtained by Schrödinger's wave mechanics and Heisenberg's matrix mechanics, which seemed to have nothing in common either in physical assumptions or in mathematical treatment, was explained by Schrödinger in one of his subsequent papers. He succeeded in proving that, unbelievable as it seemed at first, his wave mechanics was mathematically identical with Heisenberg's matrix mechanics, and that, in fact, one could derive either from the other. It was just as surprising as the statement that whales and dolphins are not fish like sharks or herring but animals like elephants or horses! But it was a fact, and today one uses wave-and-matrix-mechanics intermittently depending on one's taste and convenience. In particular, in the calculation of radiation intensities one uses matrix elements calculated on the basis of wave mechanics.

DISCARDING CLASSICAL LINEAR TRAJECTORIES

But in spite of the fact that the new Quantum Theory, either in wave-or-matrix form, gave a perfect mathematical description of atomic phenomena, it failed to illuminate the physical picture. What physical meaning should be ascribed to these mysterious waves, to these baffling matrices? How are they related to our common sense notions about matter and the world we live in? The answer to this question was given by Heisenberg in a paper published in 1927. Heisenberg starts his argument by a reference to Einstein's Theory of Relativity which, at the time of its publication (and even in some cases today) was considered contradictory to common sense by many eminent physicists. What is "common sense"? The famous German philosopher Immanuel Kant (with whose works the author is only faintly familiar) would probably define it in this way: " 'Common sense'? Why, common sense is the way things should be." And then, if he were asked: "What does 'the way things should be' mean?" he might reply: "Well, it means 'as they always were.' "§

Einstein was probably the first to realize the important fact that the basic notions and the laws of nature, however well established, were valid only within the limits of observation and did not necessarily hold beyond them. For people of the ancient cultures the Earth was flat, but it certainly was not for Magellan, nor is it for modern astronauts. The basic physical notions of space, time, and motion were well established and subject to common sense until science advanced beyond the limits that confined scientists of the past. Then arose a drastic contradiction, mainly due to Michelson's experiments concerning the speed of light, which forced Einstein to abandon the old "common sense" ideas of the reckoning of time, the measurement of distance and mechanics, and led to the formulation of the "non-commonsensical" Theory of Relativity. It turned out that for very high velocities, very large distances, and very long periods of time, things were not as they "should have been."

Heisenberg argued that the same situation existed in the field of the Quantum Theory, and he proceeded to

§ This imaginary conversation is purely in the mind of the author and not attributable to Immanuel Kant.

find out what goes wrong with the classical mechanics of material particles when we intrude into the field of atomic phenomena. Just as Einstein started the critical analysis of the failure of classical physics in the relativistic field, by the criticism of such basic notions as the *simultaneity* of two events at a distance, Heisenberg attacked the basic notion of classical mechanics, the notion of the *trajectory* of a moving material body. Trajectory had been defined from time immemorial as a path along which a body moves through space. In the limiting case used in mathematical calculations, the "body" was a mathematical point (having no dimension as Euclid defined it), while the "path" was a mathematical line (having no thickness according to the same authority). Nobody doubted that this limiting case was the best possible description of the motion, and that, by decreasing experimental errors of the coordinates and velocity of the moving particle, we could come closer and closer to exact description of the motion.

Heisenberg demurred. While the statement would undoubtedly be true if the world were governed by the laws of classical physics, he pointed out, the existence of quantum phenomena may reverse the situation. Consider first an ideal experiment in which one tries to determine the trajectory of a moving mass particle, let us say in the gravitational field of the Earth. For this purpose we build a chamber and pump the air out of it until no single molecule remains inside it (Fig. 24). On the wall of that chamber we install a little cannon C which shoots a shell with the mass m and the velocity v, let us say in a horizontal direction. On the opposite wall of the chamber is placed a small theodolite T which can be aimed at the falling particle following its path. The chamber is illuminated by an electric bulb B at the ceiling. The light from the bulb is reflected

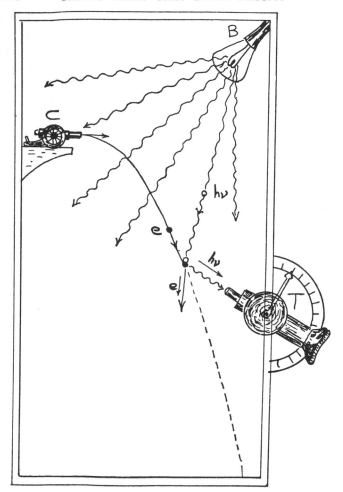

Fig. 24. Heisenberg's ideal quantum microscope inter-preting the uncertainty relations: $\Delta p \Delta q \gtrsim h$.

from the falling particle and enters the theodolite tube, and the position of the particle is indicated either on the retina of the observer's eye or on a photographic plate.

Since this is an ideal experiment we are performing,

we must take into account every possible effect that could disturb the motion of the particle, and indeed we find one, even though the air is completely evacuated. As the light from the bulb is reflected into the tube of the theodolite, it exerts on the particle a certain pressure which would deflect the particle from its expected parabolic trajectory. Can one make this disturbance negligibly small, infinitely small?

Let us proceed step by step and estimate first only ten positions of the particle; we flash the bulb only ten times during the time of fall and thus eliminate the effect of light pressure while the particle is not being looked at. Suppose that at the first trial the effect of ten kicks caused by the reflected light deflects the particle too far from the expected trajectory. There is an easy remedy. We can reduce the intensity by a necessary factor since in classical physics there is no lower limit to the amount of radiant energy that can be emitted in one flash, and also no limit to the sensitivity of the receptor of the reflected light. By cutting the intensity we can make the total disturbance during the particle's flight smaller than any small number ϵ we want to choose. If now we decide to increase the number of observed positions tenfold for a more accurate definition of the trajectory we have to flash the bulb one hundred times during the flight. The effect of radiation pressure during the total flight will increase correspondingly, and the total disturbance may become larger than ϵ. To remedy the situation we use a bulb ten times weaker and a receptor ten times more sensitive. The next steps will be to make 1000, 10,000, 100,-000, etc., observations, using correspondingly weaker bulbs and a more sensitive detector. At the limit we get an infinite number of observations without disturbing the trajectory more than by ϵ. There is another point to be considered. No matter how small the moving point may be, its optical image on the screen cannot, because of diffraction phenomena, be smaller than the wavelength λ of the light used. This disability can be remedied again by decreasing λ and using instead of visible light the ultraviolet light, X-ray, and harder and harder γ-rays. Since in classical physics there is no lower limit for the length of electromagnetic waves, the diameter of each diffraction pattern thus can be made as small as desired. Proceeding in this way, we can observe a path as thin as desired without disturbing the total motion by more than ε. Thus, within the boundaries of classical physics we can build conceptually the notion of the trajectory as a line in the Euclidean sense of the word.

But has this Euclidean state of affairs any correspondence to reality? No, said Heisenberg. He contended that the procedure of our ideal experiment becomes impossible because of the existence of light quanta. In fact, the smallest amount of energy carried by a "flash of light" is equal to $h\nu$, which corresponds to mechanical momentum $h\nu/c$. In the reflection of the flashes to the theodolite a part of this momentum will be communicated to the particle, changing its momentum by:

$$\Delta p \cong \frac{h\nu}{c}$$

Thus the number of observations increases the disturbance of the trajectory beyond any limit, and instead of moving along the parabola the particle will execute a Brownian motion, being thrown to and fro in all directions through the chamber. The only way to decrease this disturbance is to decrease ν which, because of the relation $\nu = c/\lambda$ would mean the increase of the wavelength until it became as large as the chamber. Then, instead of seeing little sparks jumping all over the screen, we would observe a system of large, overlapping diffraction spots that would cover the entire screen. Thus nothing that resembles mathematical lines can be obtained by this method.

The only alternative is to look for a compromise. We must use photons with a frequency not exceedingly high and a wavelength not exceedingly long. Since the uncertainty Δq in our knowledge of the position of the particle is $\cong \lambda = c/\nu$, we obtain:

$$\Delta p \cong \frac{h\nu}{c} = \frac{h}{\lambda}$$

or

$$\Delta p \Delta q \cong h\P$$

which is the famous Heisenberg Uncertainty Relation. In terms of velocity this relation becomes:

$$\Delta v \Delta q \cong \frac{h}{m}$$

indicating that deviations from classical mechanics become important only for very small masses. For a particle of 1 mg†† we get:

$$\Delta v \Delta q \cong \frac{10^{-27}}{10^{-3}} = 10^{-24}$$

which can, for example, be satisfied by taking:

$$\Delta v \cong 10^{-12}\frac{cm}{sec}; \ \Delta q = 10^{-12}cm$$

Thus, by this relation, the velocity of a small buckshot is uncertain within 0.3 meters per century, and the uncertainty of its position is comparable to the diameter of the atomic nucleus. Clearly, nobody will care about

¶ This relationship is often given as $\cong$ meaning "approximately equal to," or $\sim$ meaning "of the order of," or, as in Fig. 25, as $\gtrsim$ meaning "greater than or equal to."
†† A milligram (mg) is one thousandth of a gram, i.e., the mass of one cubic millimeter of water at 4°C.

such uncertainties! On the other hand, for an electron
with the mass 10^{-27} gm we have

$$\Delta v \Delta q \cong \frac{10^{-27}}{10^{-27}} \cong 1$$

Since the statement that an electron is within the atom
means that $\Delta q \cong 10^{-8}$ cm, we find that the uncertainty
of its velocity is

$$\Delta v = \frac{1}{10^{-8}} = 10^8 \frac{cm}{sec}$$

The uncertainty of the kinetic energy corresponding
to this uncertainty of velocity is

$$\Delta K \cong mv\Delta v \cong 10^{-27} \cdot 10^{-8} \cdot 10^{-8} \cong$$
$$10^{-11} \text{ erg} \cong 10 \text{ e.v.}$$

which is comparable to the total binding energy of an
electron in the atom. Obviously, in these circumstances
it is nonsensical to draw electron orbits in the atom
along lines, since the width of these lines will be comparable to the diameters of Bohr's quantum orbits!

One can say that the trajectory of a particle may
be observed not by the optical method, which leads
to the described difficulties, but by using mechanical
gadgets of some kind scattered through space and to
register the passage of particles close to them, perhaps
"jingle bells" which begin to tinkle when hit by the
passing particle. But the trouble emerges here again.
Suppose the registering gadget consists of a movable
particle within an enclosure of radius l. This particle,
being quantized, has a sequence of quantum states
differing in their mechanical momenta by the quantities

$$\Delta p \cong \frac{h}{l}$$

Thus, if the impact of the incident particle brings the

device from one quantum state to another, the incident particle loses a substantial part of its momentum. But l is the uncertainty Δq in the position of the incident particle since it can hit the registering gadget at any point on its surface. Thus we get here again

$$\Delta p \Delta q \cong h$$

for the measurements carried out by mechanical methods. It may be noted here that this jingle bell method is widely used in experimental nuclear physics under the name of the cloud chamber, where the ionized atoms of gas (with water vapor condensed on them) form long tracks showing the motion of various atomic particles. But cloud chamber tracks are not mathematical lines and are, in fact, much thicker than would be permitted by the uncertainty relation.

Since in atomic and nuclear physics the notion of classical linear trajectories inevitably fails, it is apparently necessary to devise another method for describing the motion of the material particles, and here the ψ-functions come to our aid. They do not represent any physical reality. The de Broglie waves have no mass such as we find in the case of electromagnetic waves, and whereas, in principle, one can buy half a pound of red light, there does not exist in the world an ounce of de Broglie waves. They are no more material than the linear trajectories of classical mechanics, and, in fact, can be described as "widened mathematical lines." They guide the motion of particles in quantum mechanics in the same sense as the linear trajectories guide the motion of particles in classical mechanics. But, just as we do not consider the orbits of planets in the Solar System as some kind of railroad tracks that force Venus and Mars and our own Earth to move along elliptical orbits, we may not consider the wave-mechanical continuous functions as some field of forces which influences the motion of electrons. The de Broglie-Schrödinger wave functions (or, rather, the square of their absolute values, i.e., $|\psi|^2$) just determine the *probability* that the particle will be found in one or another part of space and will move with one or another velocity.

We cannot finish this chapter without describing an exciting argument between Niels Bohr, who was the great promoter of the uncertainty relations, and Albert Einstein, who until his death remained an ardent opponent of them. This incident took place in Brussels at the Sixth Solvay Congress (1930), which was dedicated to the problems of the Quantum Theory, and it involved (as one might expect from Einstein's presence!) the uncertainty relation in *four* coordinates.

So far in this book we have written the relation $\Delta p \Delta q \cong h$ for any single coordinate and the corresponding mechanical momentum. In a three-dimensional Cartesian coordinate system, however, there are three independent relations:

$$\Delta p_x \Delta x \cong h$$
$$\Delta p_y \Delta y \cong h$$
$$\Delta p_z \Delta z \cong h$$

Since in the Theory of Relativity *time* (in the form $ct\ddagger$) serves as the fourth coordinate, and *energy* (in the form E/c) as the fourth component of mechanical momentum, one might expect that there would exist a fourth uncertainty relation

$$\Delta E \Delta t \cong h$$

and it was this topic that led to the incident at the Congress.

Einstein stepped forward to announce that he could

‡‡ Where $c = 3 \cdot 10^{10} \frac{cm}{sec}$ is the velocity of light in vacuum.

TOP *W. Heisenberg playing it cool. (Photographer unknown)*
BOTTOM *E. Fermi playing it hot. (Photographed by the author)*

propose an ideal experiment to contradict this fourth
relation. He was thinking, he said, of a box inlaid with
ideal mirrors (like Jeans' box discussed in Chapter I)
and filled with a certain amount of radiant energy. In
one of the walls was some kind of idealized photographic shutter connected with an idealized alarm
clock, which could be set so that the shutter would
operate at any given time after the box was filled with
radiation (Fig. 25). Since the clock was inside the box,
and the shutter closed, the interior of the box was completely isolated from the outside world. Einstein proposed to weigh the box prior to the alarm clock action;

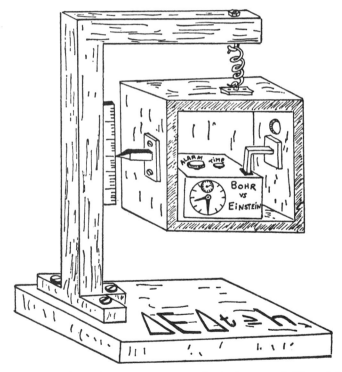

*Fig. 25. Bohr's ideal experiment which disproved Einstein's
statement that the relation $\Delta E \Delta t \gtrsim h$ is wrong.*

the weighing could be done with any desired precision, given enough time. The shutter would open at the exact moment for which the alarm clock was set, and a certain amount of radiant energy E would be let out. After the shutter closed one could again weigh the box with any desired precision. The change of the mass of the box $M_2 - M_1$ could be obtained with any precision from the two weight measurements, and being multiplied by c^2 would give the exact amount of emitted energy, so that $\Delta E = 0$. On the other hand, the ideal clock would work perfectly, so that there would be no uncertainty in the time when this energy was emitted, and $\Delta t = 0$, too. Having $\Delta E = 0$ along with $\Delta t = 0$ would kill the fourth uncertainty relation.

The argument seemed very persuasive, and Bohr had nothing to say. But the next morning, after an almost sleepless night, Bohr, his face radiant, appeared in the meeting hall with an explanation. In order to weigh the box, he pointed out, one should permit it to move in a vertical direction no matter whether balance scales or spring scales were used. The clock, changing its position in the gravitational field of the Earth, would lose or gain time according to Einstein's law concerning the influence of the gravitational potential on the rate of the clock. An uncertainty Δt in the time of the shutter release would be introduced. On the other hand, the amplitude of the vertical oscillations of the box, which determines Δt, is connected through the ordinary relation $\Delta p_z \cdot \Delta z \cong h$ with the mass change which sets the box swinging when energy is lost. Juggling the equations, Bohr easily came to the conclusion that $\Delta E \Delta t \cong h$, thus defeating Einstein's argument by the use of Einstein's own most important discoveries.

This chapter has concentrated on Heisenberg's Principle, rather than on his personal characteristics, but the author can add that Heisenberg was an expert skier, played ping-pong with the left hand, and, in spite of his great fame as a physicist, was better known in Leipzig (where he was a professor) as a first-class piano player.

CHAPTER VI

P. A. M. DIRAC AND ANTI-PARTICLES

The Theory of Relativity and the Quantum Theory, which appeared almost simultaneously at the start of the present century, were two great explosions of the human mind and shook the very foundations of classical physics: relativity in the case of velocities approaching that of light; quantum in the case of motions confined to very small (atomic) dimensions. But for almost three decades these two great theories stood apart, more or less, from each other. Bohr's original theory of quantized orbits, as well as Schrödinger's wave equations into which it developed, were essentially non-relativistic; both were applicable only to particles moving with a velocity small as compared with that of light. But the velocities of electrons within the atoms are not that small. For example, the electron on the first hydrogen orbit, calculated on the basis of Bohr's theory, has the velocity of 2.2×10^8 cm/sec, which is only a little less than 1 per cent of the speed of light. The velocity of electrons inside heavier atoms is considerably larger. Of course, a few per cents is not too much, and the calculated value could be improved by introducing "relativistic corrections," which would make the agreement with direct experimental measurement somewhat better. But this was only an improvement and not the completion of the theory.

Another trouble arose in the case of the electron's magnetic moment. In 1925, Goudsmit and Uhlenbeck showed that in order to explain certain details of atomic spectra it is necessary to ascribe to the electron certain angular and magnetic momenta[†] commonly known as the *electron spin*. The naïve picture at that time was that the electron is a little charged sphere about 3×10^{-13} cm in diameter. The rapid rotation of this sphere around its axis was supposed to produce a magnetic moment, resulting in additional interaction with its orbital motion and with the magnetic moments of other electrons. It turned out, however, that in order to produce the necessary magnetic field the electron would have to rotate so fast that the points on its equator would move at much higher velocities than light! Here again one encountered a conflict between quantum and relativistic physics. It was becoming clear that relativity and quantum physics could not just be added together. A more general theory which would contain both relativistic and quantum ideas in a harmoniously unified form was needed.

[†] See Francis Bitter, *Magnets*. Doubleday, Science Study Series (1959).

The most important step in this direction was taken in 1928 by a British physicist, P. A. M. Dirac, who started his career as an electrical engineer but, finding it difficult to get satisfactory employment, applied for a fellowship in physics at Cambridge University. His application (which was accepted) now hangs, attractively framed, in the University Library, side by side with the Nobel Prize certificate which he received not too many years after changing from electrical engineering to quantum physics.

Now it often happens that "absent-minded-professor" stories grow up around famous scientists. In most cases these stories are not true, merely the inventions of wags, but in the case of Dirac all the stories are really true, at least in the opinion of this writer. For the benefit of future historians we give some of them here.

Being a great theoretical physicist, Dirac liked to theorize about all the problems of daily life, rather than to find solutions by direct experiment. Once, at a party in Copenhagen, he proposed a theory according to which there must be a certain distance at which a woman's face looks its best. He argued that at $d = \infty$ one cannot see anything anyway, while at $d = 0$ the oval of the face is deformed because of the small aperture of the human eye, and many other imperfections (such as small wrinkles) become exaggerated. Thus there is a certain optimum distance at which the face looks its best.

"Tell me, Paul," I asked, "how close have you ever seen a woman's face?"

"Oh," replied Dirac, holding his palms about two feet apart, "about that close."

Several years later Dirac married "Wigner's sister," so known among the physicists because she was the sister of the noted Hungarian theoretical physicist Eugene Wigner. When one of Dirac's old friends, who had not yet heard of the marriage, dropped into his home he found with Dirac an attractive woman who served tea and then sat down comfortably on a sofa. "How do you do!" said the friend, wondering who the woman might be. "Oh!" exclaimed Dirac, "I am sorry. I forgot to introduce you. This is . . . this is Wigner's sister."‡

Dirac's sense of quantum humor was often demonstrated at scientific meetings. Once, in Copenhagen, Klein and Nishina reported their derivation of the famous Kline-Nishina formula describing collisions between electrons and gamma quanta. After the final formula was written on the board, somebody in the audience who already had seen the manuscript of the paper remarked that in the formula as written on the blackboard the second term had the negative sign, whereas in the manuscript the sign was positive.

"Oh," said Nishina, who was delivering the talk, "in the manuscript the signs are certainly correct, but here on the blackboard I must have made a sign mistake in some place."

"In *odd* number of places!" commented Dirac.

Another example of Dirac's acute observation has a literary flavor. His friend Peter Kapitza, the Russian physicist, gave him an English translation of Dostoevski's *Crime and Punishment*.

"Well, how do you like it?" asked Kapitza when Dirac returned the book.

"It is nice," said Dirac, "but in one of the chapters the author made a mistake. He describes the Sun as

‡ In a recent conversation with Mrs. Dirac (in Austin, Texas, of all places!) the author asked whether this story is really true. She said that what Dirac actually said was: "This is Wigner's sister, who is now my wife."

rising twice on the same day." This was his one and only comment on Dostoevski's novel.§

Another time, visiting Kapitza's home, Dirac was watching Anya Kapitza knitting while he was talking physics to Peter. A couple of hours after he left, Dirac rushed back, very excited. "You know, Anya," he said, "watching the way you were making this sweater I got interested in the topological aspect of the problem. I found that there is another way of doing it and that there are only two possible ways. One is the one you were using; another is like that. . . ." And he demonstrated the other way, using his long thin fingers. His newly discovered "other way," Anya informed him, is well known to women and is none other than "purling."

Just to finish "Dirac stories" before we go on to his scientific achievements, let me mention one more. At the question period after a Dirac lecture at the University of Toronto, somebody in the audience remarked: "Professor Dirac, I do not understand how you derived the formula on the top left side of the blackboard."

"This is not a question," snapped Dirac, "it is a statement. Next question, please."

UNIFYING RELATIVITY AND QUANTUM THEORY

Let us turn now to Dirac's achievement in physics. As was stated in the beginning of the chapter, the Quantum Theory and the Theory of Relativity could not be fitted exactly together like pieces of a Chinese puzzle. One can get a very close fit, but there were always some minor discrepancies, making the solution not quite per-

§ The author, who heard this from Kapitza, was too lazy to read *Crime and Punishment* once more to find out in which chapter this occurred. But some of the readers of this present book may wish to try.

fect. Schrödinger's wave equation of the Quantum Theory looked very similar to the classical wave equation describing the propagation of sound or electromagnetic waves, but . . .

In classical physics the quantities under consideration, be they the density of air or electromagnetic forces, always enter the wave equation in the form of *second derivatives;*¶ that is, the rates of rates of change on x, y, z, and t, conventionally written as

$$\frac{\partial^2 u}{\partial x^2}; \frac{\partial^2 u}{\partial y^2}; \frac{\partial^2 u}{\partial z^2}; \text{ and } \frac{\partial^2 u}{\partial t^2}$$

The exact mathematical solution of such equations always leads to harmonic waves propagating through space. Schrödinger's wave equation contained the second derivatives on x, y, and z, but only the first derivative on t. The reason was that this equation was derived from classical Newtonian mechanics in which the acceleration of a moving material particle is proportional to the acting force. In fact, if x is the position of the particle, its velocity v (that is, the rate of change of its position with time) is the *first* derivative of x on t

$$\left(\frac{\partial x}{\partial t}\right)$$

whereas its *acceleration* a (that is, the rate of change of its velocity with time) is the second derivative:

$$\frac{\partial\left(\dfrac{\partial x}{\partial t}\right)}{\partial t}$$

¶ The notion of derivatives is discussed in an elementary way in Chapter 3 ("Calculus") of the author's book *Gravity*, published in 1962 in this same series. See also *Mathematical Aspects of Physics* (Doubleday, Science Study Series, 1963) by Francis Bitter.

customarily written as

$$\frac{\partial^2 x}{\partial t^2}$$

On the other hand, the force F is the first derivative of the potential P on the position:

$$\frac{\partial P}{\partial x}; \frac{\partial P}{\partial y} \text{ and } \frac{\partial P}{\partial z}$$

Thus, the basic Newton law of motion, stating that the acceleration is proportional to the force, contained the first derivatives on the space coordinates, and the second derivative on time. This fact made the Newtonian equation of the motion of a particle mathematically inhomogeneous, giving to the time t a different status than to the coordinates x, y, z. This situation, which existed for centuries in classical mechanics, is reflected in Schrödinger's non-relativistic wave mechanics, in which space and time are treated as quite different entities.

But as soon as we try to formulate the laws of Quantum Theory on a relativistic basis, we run into the difficulty that space and time are much more closely connected with each other. In fact, following up Einstein's basic ideas, H. Minkowski formulated the notion of a four-dimensional space-time continuum in which time, multiplied by an imaginary unit $i = \sqrt{-1}$, is regarded as equivalent to the three space coordinates. In Minkowski's world there is no difference between $x, y, z,$ and ict (where c†† is introduced through purely dimensional considerations).

(In this book, dedicated to the Quantum Theory, we have no space to discuss the Theory of Relativity in detail; the reader unfamiliar with the subject must get

†† And the factor c (velocity of light) to keep the physical dimensions correct.

his information from other books.‡‡ The author must assume, however, that the person reading the following chapters has at least an elementary acquaintance with the basic ideas of Einstein's theory.)

As was discussed earlier, the wave mechanical equation must contain the same derivatives in all four coordinates. Schrödinger's equation, however, being derived from Newton's equation, does not satisfy that condition. The first attempts to straighten out this defect were made independently by O. Klein and W. Gordon, who turned Schrödinger's non-relativistic wave equation into a relativistic form simply by introducing the second derivatives on time, instead of the first derivatives. But although the Klein-Gordon wave equation looked very nice and very relativistic, it suffered from a number of internal contradictions, and all attempts to introduce the electron spin into it in any reasonable way led to a complete failure.

Then, one evening of the year 1928, sitting in an armchair in his St. John's College study, with his long legs stretched toward the burning logs in the fireplace, Paul Adrien Maurice Dirac hit on a very simple and very brilliant idea.

If no good result can be obtained by using the second derivatives on the time coordinate in the relativistic wave equation, why not use the first derivatives on space coordinates in it? Of course, it would mean the introduction of more imaginary units i, but it would make the wave equation symmetrical in space and time. Thus appeared Dirac's linear (containing only first derivatives) equation which, applied to the hydrogen atom, immediately led to glorious results. All the splittings of the spectral lines, which had stubbornly resisted interpretation in terms of the spin and magnetic

‡‡ See Hermann Bondi, *Relativity and Common Sense*. Doubleday, Science Study Series (1964).

moment of the electron, came out completely correct on the basis of the new theory. This success was particularly surprising because, in formulating his equation, Dirac aimed only to make it relativistically correct; the spinning electron appeared as the bonus for uniting in a correct way the relativity and quantum theories. And it was not a small electrically charged and rapidly rotating sphere but a point charge which, by virtue of Dirac's equation, behaved *as if it were* a tiny magnet.

But, having written the wave equation, which represented a perfect unification of the relativity and quantum theories, Dirac had to face another difficulty which was characteristic of any attempt at uniting these two theories. According to the famous Einstein relation, a rest mass m_0 (expressed in grams) was equivalent to the energy $m_0 c^2$ (expressed in ergs) where c is the velocity of light. If that mass is moving with a certain velocity v, thus having (in the first approximation) the kinetic energy $K = \dfrac{1}{2} m_0 v^2$§§ the total energy is:

$$E = \frac{m_0 c^2}{\sqrt{1 - \dfrac{v^2}{c^2}}} \cong \P\P\, m_0 c^2 + \frac{1}{2} m_0 v^2$$

But, due to the mathematical properties of Einstein's relativistic mechanics, one should expect also the type of motion corresponding to the total energy:

$$E = -\frac{m_0 c^2}{\sqrt{1 - \dfrac{v^2}{c^2}}} \cong \P\P\, -m_0 c^2 - \frac{1}{2} m_0 v^2$$

§§ or more exactly: $m_0 c^2 \left(\dfrac{1}{\sqrt{1 - \dfrac{v^2}{c^2}}} \right) - 1$

which becomes equal to $\frac{1}{2} m_0 v^2$ if $v \ll c$.

¶¶ for $v \ll c$.

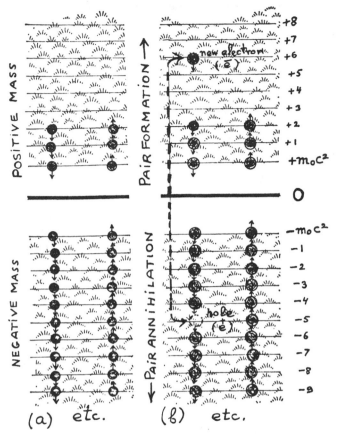

Fig. 26. Dirac's picture of the energy level distribution of the particles with positive and negative mass. On the left (a) all negative energy levels are completely filled up, and only six ordinary electrons can exist on the normal positive levels. On the right (b) one of the electrons from a negative level is lifted to a positive level, leaving behind it a "hole" which behaves as an ordinary positive electron with positive mass. If this extra electron from a positive level falls back into the hole (annihilation process of $\bar{e}$ and $\grave{e}$ the energy difference will be emitted as γ-radiation.

This equation can be obtained from the previous one by writing $-m_0$ instead of $+m_0$, which means physically the introduction of *negative mass*. Thus relativistic mechanics permits in principle two separate sets of levels: those with rest energy $+m_0c^2$ and higher, another with the rest energy $-m_0c^2$ and lower (Fig. 26).

While the energy levels shown in the upper part of the diagram $(E>0)$ correspond to familiar types of motion of material particles (electron, proton, etc.), the energy levels in the lower part of the diagram $(E<0)$ do not correspond to any physical reality. Particles having negative inertial mass do not correspond to anything observed in nature. Indeed, because of the negative value of their mass they would be accelerated in the direction *opposite* to the force acting on them, and, in order to stop a moving particle of that kind, one should push it in the direction of its motion and not against it! Imagine two particles, let us say two electrons, with numerically equal masses having, however, opposite signs $(+m$ and $-m)$. According to the Coulomb law, they will be repelled by each other by electrostatic forces having the same numerical values, but acting in the opposite direction. If both particles had positive masses, this interaction would result in equal but oppositely directed accelerations (Fig. 27a), and they would fly away from each other with increasing velocities. If, however, one of the particles has a negative mass (Fig. 27b), it will be accelerated in the same direction as the other particle, and they will fly together, keeping a constant distance between themselves and speeding up beyond any limit ($<c$ of course). There is no contradiction of the Law of Conservation of Energy, since the combined kinetic energy of the two particles will be:

$$\frac{1}{2}mv^2 + \frac{1}{2}(-m)\,v^2 = 0$$

Fig. 27. Interaction between the particles with a positive and those with a negative mass.

the same as it was before the motion started. The whole thing is completely fantastic, and particles having these properties have never been observed.

In classical relativistic mechanics (which does not take into consideration the quantum phenomena) the difficulty of particles with negative mass can easily be eliminated. Indeed, as one can see from Fig. 26, the regions of positive and negative energies are separated by an interval of 2 m_0c^2 (about 1 million electron volts in the case of electrons). Since in non-quantum mechanics (classical and relativistic) the changes of energy must be continuous, a particle from the upper part of the diagram cannot move to the lower part because it would require a discontinuous change of its energy. Thus in the physical description of nature the states with negative mass could be rejected as undesirable mathematical possibilities. However, as soon as one introduces the quantum phenomena, the situation changes quite radically: in Quantum Theory electrons and other elementary particles just love to *jump* from higher levels to lower ones. Thus in the relativistic Quantum Theory a paradoxical event should happen: all the normal electrons would jump from positive mass states to negative mass states, and all the Universe would go haywire!

The only way Dirac could see to overcome this paradox was by the use of the Pauli Principle, and the assumption that all the states corresponding to the negative mass are already occupied (two electrons with opposite spin per each state), leaving no place for electrons from the positive mass states to jump into. The situation is similar to that of the familiar electron-shells of an atom where the electron from M shell cannot jump down to L or K shells because these are completely occupied by the electrons which got there first. But, whereas atoms are limited structures containing a finite number of electrons, Dirac's theory pertained to limitless space and called for an infinite number of electrons per each cubic centimeter of vacuum. So far so good, if one neglects the infinite mass of these electrons which, according to Einstein's relativistic theory of gravity (sometimes called the General Theory of Relativity), would make the radius of curvature of empty space equal to zero!

Putting aside this difficulty, Dirac asked himself whether that distribution of negatively charged electrons with negative mass would be observable, that is, detectable with any kind of physical measuring instruments. The answer was no. With no kind of electric equipment can one detect a uniform charge distribution in space, no matter how high it is per unit volume. To understand that statement, let us imagine a deep sea fish which never comes up to the ocean's surface and never sinks all the way down to the ocean floor. If we assume that the ocean water is frictionless (something like liquid helium), we must conclude that the fish, intelligent as it may be, cannot tell whether it is moving through water or through a complete vacuum. And, if something is unobservable, it should not be used in the physical description of nature. Our deep sea fish is accustomed to seeing objects moving downward, be they bits of garbage thrown overboard from ships sailing across the ocean or, in rare cases, sinking ships themselves. Thus, following Aristotle, the fish will conceive the notion of gravity which makes all material objects move downward.

But suppose now that a sinking, empty Coca-Cola bottle, or a sinking ocean liner, has some trapped air which is released when the vessel hits the bottom. What will our intelligent fish see? It will see a bunch of silvery spheres (air bubbles in common human language) rising upward. What will our intelligent fish think on observing these objects? Well, it will be astonished that they move in the direction opposite to that of the force of gravity and it will be inclined to ascribe to them a mass of the opposite sign in respect to ordinary objects, which move downward.

A somewhat closer analogy may be given by considering a complex atom with completed K, L, and M shells, which has been hit by a hard X-ray and lost one of the two electrons on the K shell. There will be an empty space (Pauli vacancy) in the K shell, and one of the L shell electrons will jump into it, leaving an empty space in its shell. The next step will be transition of a most agile electron from the M shell into the vacant place in the L shell. Of course, there is also the possibility that the M shell will beat the L shell to it, and the vacancy in the K shell will be occupied directly by one of the M shell electrons.

ANTI-PARTICLE PHYSICS

But we can look on the problem from a different point of view. The fact that a negative electron is missing from the K shell is equivalent to a positive charge sitting there. The transition of a negative electron from the L shell down to the K shell is equivalent to the raising of that positive charge from the K shell up to the L shell and later up to the M shell. From this point of view we have a fictitious positively charged particle moving from the lowest K level to the much higher M level and later out into interatomic space. Since, according to the Coulomb law, the positively charged nucleus should repel the positively charged fictitious electron, everything is fine and dandy.

Returning to Dirac's ocean filled with negatively charged electrons possessing negative mass, we can ask ourselves how an experimentalist will apprehend the situation in which one negative electron with negative mass is missing from its level. Apparently there are two straightforward conclusions: (1) The absence of negative charge will be observed as the presence of positive charge. Thus the experimentalist will observe a particle with electric charge $+e$. (2) The absence of negative mass is equivalent to the presence of positive mass. Thus the particle will behave in a normal way and will be observed as a positively charged ordinary particle. Having gone so far, Dirac overstretched his idea. He thought that it could be proved that the numerical value of the mass of this hole in the ocean of electrons with negative mass is equal to about 1840 times the mass of an ordinary electron. If this were really so, the holes in Dirac's ocean would be observed as ordinary protons.

Dirac's paper published in 1930 (or rather, the private conversations and correspondence prior to its publication) resulted in violent opposition to his idea. Niels Bohr, who for some reason unknown to the writer was interested in elephants, composed a hunting story "How to Catch Elephants Alive." For the benefit of African big game hunters he proposed the following

TOP   *N. Bohr and A. Einstein, presumably during the 1930 Solvay Congress in Brussels. (Photographer unknown)*
BOTTOM   *George Gamow on a mountaintop discussing nuclear physics with Leon Rosenfeld. (Photographed by Rudolph Peierls)*

method: At a watering spot on a river where the elephants come to drink and to wash, one should erect a large poster explaining in a short sentence Dirac's proposal. When the elephant, who is proverbially a very wise animal, comes to have a drink of water, he reads the text on the poster and becomes spellbound for several minutes. Using this time interval, the hunters hiding in the bush will slip out and tie the elephant's legs securely with heavy ropes. Then the elephant is shipped to the Hagenbeck Zoo in Hamburg.

Pauli, who liked jokes too, made some calculations which showed that if protons in the hydrogen atoms were Dirac's holes the electrons would jump into them within a negligible fraction of a second, and the hydrogen atom (as well as the atoms of all other elements) would be annihilated instantaneously in a burst of high-frequency radiation. Pauli proposed what was known as the "Second Pauli Principle," according to which any theory suggested by a theoretician would become immediately applicable to his body. Thus Dirac would be turned into gamma-rays before he could tell anybody about his idea.

All this was a lot of fun, but one year later, after the publication of Dirac's paper, an American physicist, Carl Anderson, studying cosmic ray electrons passing through a strong magnetic field, found that while one-half of them were deflected in the direction expected for properly behaving negatively charged particles, the other half were deflected at the same angle *in the opposite direction*. Those were the positively charged electrons, sometimes called positrons, predicted by Dirac's theory. Experimental studies of the positrons have shown that they behave exactly as Dirac's holes were supposed to do. Although the positrons were first discovered in cosmic rays, one soon found that they could also be produced under controlled laboratory conditions simply by shooting hard gamma-rays at the metal plates. Colliding with the atomic nucleus, a γ-quantum disappears and all its energy is converted into two electrons, one negative and one positive, as is shown in Fig. 28a. Since the mass of one electron

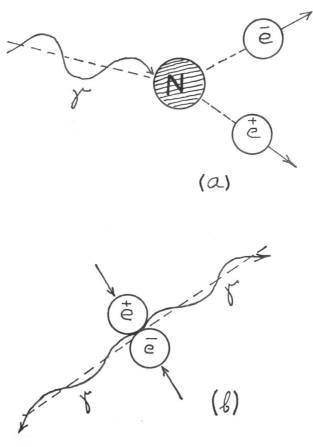

(a)

(b)

Fig. 28. The "creation" and "annihilation" of negative and positive electrons (ē and ė) according to Dirac's theory: (a) a high-energy γ-ray hits a nucleus (N) and turns into ē and ė pair; (b) an ē and ė pair collide in free space and produce two γ-rays moving in opposite directions.

expressed in energy units is equal to 0.5 mev, the process takes place only if the energy $h\nu$ of the γ-ray is greater than 1.0 mev. The excess of energy:

$$h\nu - 2m_0c^2$$

is communicated to the electron pair $(\overset{.}{e}, \bar{e})$ "created" in the collision. The fates of these two electrons are quite different. The negative (ordinary) electron $\bar{e}$ is slowed down gradually in collisions with other negative electrons forming matter and becomes one of them. The positive electron $\overset{.}{e}$ does not last long but is annihilated in the collision with one of the (ordinary) negative electrons emitting two γ-quanta (Fig. 28b). The terms "created" and "annihilated" should not be understood in a metaphysical sense; one can just as well say that ice is "created" from water when it is brought below the freezing point, and "annihilated" at room temperature, turning into water. The Laws of Conservation of Mass and Energy (which are actually one single law because of Einstein's formula $E = mc^2$) are sustained in both processes, and we deal here just with the transformation of radiation into particles and the transformation of particles into radiation on an equal basis.

The detection of anti-electrons (positrons) raised the question whether or not there also exist anti-protons, particles having the mass of a proton but carrying a negative charge. Since a proton is some 1840 times heavier than an electron, the production of proton-anti-proton pairs demands energies of billions, rather than millions, of electron volts. With this in mind, the Atomic Energy Commission spent proportionate amounts of dollars to build accelerators which could communicate to nuclear projectiles the necessary amounts of energy. And within a few years two giant accelerators were constructed in the United States: a

Bevatron at the Radiation Laboratory of the University of California at Berkeley; and a *Cosmotron* at the Brookhaven National Laboratory on Long Island, New York. Soon thereafter similar European machines were built, at CERN near Geneva, Switzerland, and in Soviet Russia, near Moscow. It was a hard competition which was finally won by Californians when Emilio Segré and his co-workers announced in October 1955 that they had detected negative protons ejected from the bombarded targets. Later, they also found the antineutrons, the particles which are annihilated in collision with ordinary neutrons. As we shall see later in the book, all the other more recently discovered particles (various kinds of mesons and hyperons) also possess their antis.

Thus, although Dirac failed in his original intention to explain a proton as an anti-electron, he opened a broad field of anti-particle physics.

There are two unsolved mysteries about anti-particles. The atoms forming our globe are built of negative electrons, positive protons, and ordinary neutrons. According to astronomical studies the same is true of the entire planetary system and the Sun itself. In fact, the protons and electrons ejected from the Sun and entering the terrestrial atmosphere are the (ordinary) positive protons and negative electrons. More uncertain, but probably true, is the statement that all the stars and interstellar material of the Milky Way are formed of ordinary matter, since otherwise one would observe intensive gamma-radiation from all the various parts of our galaxy. But what about billions of other galaxies which are separated from our Milky Way by millions of light years? Is our Universe lopsided, being formed entirely from "ordinary" matter, or is it a collection of galaxies, half of which are composed of "ordinary" matter while the other half are composed of "anti-matter"? This we do not know, and there seems to be no way to find out.

Another riddle is whether or not the anti-particles produced in abundance in our modern accelerators have a positive or a negative gravitational mass. It seems at first sight that this question might easily be answered by direct experiment. Just produce a beam of anti-protons in a high-energy accelerator and send it horizontally along an evacuated tube and see whether under the action of terrestrial gravity it will bend down as a horizontally thrown stone or bend up. If the latter one would be justified in inferring that anti-particles are repelled by the mass of the Earth. The trouble is, however, that the anti-particles produced in our laboratories move with velocities almost equal to that of light (3×10^{10} cm/sec). Thus, if the tube is, let us say, 3 km long (3×10^5 cm), the anti-particles will pass it in the time interval of 10^{-5} sec. According to the law of free fall, they will be displaced downward (or upward, in the case of negative gravitational mass) by the amount of $\frac{1}{2} gt^2$ cm where g is about 10^3 cm/sec². If $t = 10^{-5}$ sec, the vertical displacement will be of the order of $10^3 \times 10^{-10} = 10^{-7}$ cm, which is comparable with atomic diameter! It is clear that no experimental arrangement can detect such a small deflection of the beam. To carry out the experiment one could try to slow down the anti-particle to more reasonable speeds, say a few centimeters per second, when downward or upward deflection would become easily noticeable. But how does one do it? In the atomic piles one slows down neutrons by passing them through various "moderators" (carbon, or heavy water) where the neutrons gradually lose their energy in collisions with other atoms. But we cannot do this in the case of anti-particles, since on passing through any moderator formed by ordinary matter they will be annihilated in the very first collision. Thus the question still remains unanswered.

In conclusion, it may be remarked that the proof of the negative gravitational mass of anti-particles would be quite useful for the solution of various cosmological problems. If both ordinary and anti-particles were created uniformly through the space of the Universe, gravitational attraction between the particles of the same kind, and the hypothetical gravitational repulsion between the particles of the opposite kind, would result in mutual separation. Large regions of space populated exclusively by ordinary matter would be formed and so would other regions populated exclusively by anti-matter. This separation would gratify our notion of the symmetry of nature. But we do not know, and we also do not know whether or not we will ever know.

CHAPTER VII

E. FERMI AND PARTICLE
TRANSFORMATIONS

In olden times any physicist could handle both the experimental and theoretical parts of his science. The
outstanding example is Sir Isaac Newton, who created
the Theory of Universal Gravity (inventing for that
purpose a mathematical discipline now known as the
calculus). He also carried out important experimental
studies proving that white light is formed by an overlap
of a spectrum of various colors, and he was the first to
construct a reflecting telescope. But as the field of physics grew broader and broader, the experimental techniques and mathematical methods became more and

more complicated, too much for any one man to handle. The physics profession split into two branches, "experimentalists" and "theoreticians." The great theoretician Albert Einstein never did an experiment with his own hands (to the author's knowledge, at least), while the great experimentalist Lord Rutherford was so poor in mathematics that the famous Rutherford formula for α-particle scattering was derived for him by a young mathematician, R. H. Fowler. Today, as a rule, a theoretical physicist never dares touch experimental equipment for fear of breaking it (see the Pauli Effect in Chapter III), while experimentalists are lost in the turbulent flow of mathematical computations.

Enrico Fermi, born in Rome in 1901, represented a rare example of an excellent theoretical as well as experimental physicist. One of his important contributions to theoretical physics was the study of degenerated electron gas, which had important consequences in the electron theory of metals as well as in the understanding of the super-dense stars known as white dwarfs. Another important work was the formulation of the mathematical theory of particle transformation, involving the emission of mysterious chargeless and massless particles proposed earlier by Pauli.

Fermi was a sturdy Roman boy with a great sense of humor. While he was still a professor in the University of Rome, Mussolini awarded him a title: "Eccellenza" (His Excellency). Once he had to attend a meeting of the Academy of Sciences at the Palazzo di Venezia, which was strongly guarded because Mussolini himself was to address it. All other members arrived in large foreign-made limousines driven by uniformed chauffeurs, while Fermi drew up in his little Fiat. At the gate of the Palazzo he was stopped by two *carabinieri* who crossed their weapons in front of his little car and asked his business there. According to the story he told to the author of this book, he hesitated to say to the guards: "I am His Excellency Enrico Fermi," for fear that they would not believe him. Thus, to avoid embarrassment, he said: "I am the driver of His Excellency, Signore Enrico Fermi." *"Ebbene,"* said the guards, "drive in, park, and wait for your master."

Although the idea of chargeless and massless particles accompanying electrons emitted in β-transformations was originally conceived by Pauli, Fermi was the first to develop the strict mathematical theory of β-emission coupled with the emission of Pauli's pet particles, and to show that it fits perfectly with the observed facts. He was also responsible for its present name, *neutrino*. The point is that Pauli called his protégé the *neutron,* which was all right since the particle called "neutron" today (the chargeless proton) had not then been discovered. However, that name was not "copyrighted" since it was used only in private conversations and correspondence but never in print. When, in 1932, James Chadwick proved the existence of a chargeless particle with a mass closely equal to that of a proton, he called it the *neutron* in his paper published in the *Proceedings of the Royal Society of London.* When Fermi, still being a professor in Rome, reported Chadwick's discovery at the weekly physics seminar, somebody from the audience asked whether "Chadwick's neutron" is the same as "Pauli's neutron." "No," answered Fermi (naturally speaking in Italian), *"i neutroni di Chadwick sono grandi e pesanti. I neutroni di Pauli sono piccoli e leggieri; essi debbono essere chiamati neutrino."*†

Having made this philological contribution, Fermi

---

† In Italian *neutrino* is the diminutive of *neutrone;* in other words, "the little neutron." Fermi's reply is translated: "No, the neutrons of Chadwick are large and heavy. Pauli's neutrons are small and light; they have to be called neutrinos."

set forth to develop a mathematical theory of β-transformation in which an electron (positive or negative) and a neutrino are emitted simultaneously by unstable atomic nuclei, sharing the available energy at random among themselves. He shaped his theory along lines similar to the theory of light emission by atoms, where an excited electron makes the transition to a state of lower energy, in the process liberating the excess energy in the form of a single light quantum. The motion of the electron before the discontinuous transition was described by the wave function spreading over a comparatively large area. After the transition the electron's wave function shrank to a smaller size, and the liberated energy formed a divergent electromagnetic wave spreading through the surrounding space. The forces responsible for that transformation were the familiar forces acting between the electromagnetic field and the point charge. Thus, their effect could easily be computed on the basis of the existing theory. It was found that the computed probabilities of electron transitions stood in perfect agreement with the observed intensities of spectral lines.

In his theory of β-decay Fermi faced a much more complicated situation. In this case, a neutron occupying a certain energy state within the nucleus turned into a proton, thus changing its electric charge. Also, instead of a single light quantum, two particles (an electron and a neutrino) were emitted simultaneously.

The Forces Behind β-Transformation

The main difficulty, however, was that, whereas in the case of light emission the forces governing the process were the familiar electromagnetic forces, the forces responsible for β-transformation were absolutely unknown and Fermi had to make a guess as to what they were. Characteristically for a genius, he decided to make the simplest possible assumption—that the probability of the transformation of a neutron into a proton (or vice versa), resulting in the formation of an electron (negative or positive) and a neutrino,‡ is simply proportional to the product of the intensities of four corresponding wave functions at any given point within the nucleus. The coefficient of proportionality, which Fermi designated by the letter g, had to be determined by comparison with the experimental data. Using rather complicated mathematics, Fermi was able to compute what the shape of the β-energy spectrum should be, and how the rate of β-decay should depend on the amount of energy involved if his simple interaction hypothesis was correct. The result was in brilliant agreement with the observed curves.

The only flaw in Fermi's theory of β-decay was that the numerical value of the constant g (3×10^{-14} in dimensionless units§) could not be derived from the theory and had to be taken directly from observation. The extreme smallness of the numerical value of g is responsible for the fact that, whereas emission of a γ-quantum by a nucleus occurs within 10^{-11} seconds, the emission of an electron-neutrino pair may take hours, months, or even years. This is why all particle transformations are known in modern physics as *weak interactions*. It is the task of physics of the future to explain these extremely weak interactions in all processes involving the neutrino emission absorption.

‡ We will not go here into the distinction between a neutrino and an anti-neutrino.

§ $|g| \cdot \dfrac{|mc^2|}{|\sqrt{2\pi h^3}|}$, where m is the electron mass, c the velocity of light, and h the quantum constant.

Using Fermi Interaction Laws

Analogous to the β-decay processes¶

$$n \longrightarrow p + \bar{e} + \nu + \text{energy}$$

and

$$p + \text{energy} \longrightarrow n + \dot{e} + \nu$$

are other processes which are also subject to Fermi interaction laws. One is the absorption of the atomic electron by a nucleus which is unstable in respect to positive β-decay. Instead of emitting a positive electron and a neutrino, a nucleus may absorb a negative electron from its own electron shell, emitting a neutrino according to the formula:

$$_z(\text{nucl.})^A + \bar{e} \longrightarrow _{z-1}(\text{Nucl.})^A + \nu + \text{energy}$$

Since the atomic electron absorbed by the nucleus in such a process is one of the electrons from the K-shell (nearest to the nucleus), it is usually known as "K-capture." The simplest example of such a process is the unstable isotope of beryllium, Be^7, which may transform either according to formula††

$$_4Be^7 \longrightarrow _3Li^7 + \dot{e} + \nu + \text{energy}$$

or

$$_4Be^7 + (\bar{e})_{K-shell} \longrightarrow _3Li^7 + \nu + \text{energy}$$

In the latter case, cloud chamber photographs show just a single track (that of $_3Li^7$), and the situation is similar to an incident described by H. G. Wells in his

¶ According to energy considerations, the first process occurs in the case of the free neutron, as well as in the case of neutrons bound inside the nucleus, whereas the second occurs only within the complex nuclei where the additional energy supply can be obtained from other nucleons.

†† The lower index on the left gives the atomic number whereas the upper index on the right gives atomic weight.

well-known story *The Invisible Man,* where a London constable was kicked in the pants from behind and, turning around, could not see anybody who could have kicked him. Observational studies of the K-capture processes showed that the frequency of their occurrence agrees exactly with that predicted by Fermi's theory.

Another interesting process belonging to the same category is the H-H (Hydrogen-Hydrogen) reaction, first proposed by Charles Critchfield, which is responsible for the energy production in our Sun and other fainter stars.‡‡ During the short interval of time while two colliding protons are in close contact, one of them turns into a neutron through emission of a positive electron and a neutrino, forming the nucleus of deuterium (heavy hydrogen) according to the equation:

$$_1p^1 + {}_1p^1 \longrightarrow {}_1d^2 + \acute{e} + \nu + \text{energy}$$

The probability of this process can be predicted exactly on the basis of Fermi's theory.

The last but not the least example of the Fermi interaction is the process by means of which F. Reines and C. Cowan directly proved the existence of neutrinos. It is:

$$_1p^1 + \nu + \text{energy} \longrightarrow {}_0n^1 + \acute{e}$$

Reines and Cowan observed it in a chamber placed close to an "atomic pile," at the Savannah River Atomic Energy Project. The number of observed neutrons and positive electrons formed simultaneously in the chamber subjected to extensive neutrino bombardment turned out to be exactly equal to that predicted by the Fermi theory. The interaction is so weak that,

‡‡ In the case of brighter stars, Sirius for example, the main energy-producing reaction is the so-called carbon cycle, proposed independently by C. von Weizsacker and H. Bethe.

in order to absorb one-half of the emitted neutrinos, one should use a liquid hydrogen shield several light years thick! Fermi's theory of the processes involving neutrinos also applies to many cases of decay of new elementary particles discovered during more recent years, and one speaks today about the "Generalized Fermi Interaction."

FERMI'S RESEARCH IN NUCLEAR REACTIONS

Along with his theoretical studies, Fermi was involved in extensive experimental research on nuclear reactions in heavy elements bombarded by slow neutrons, and the formation of trans-uranium elements ($z > 92$), and for this work he received the Nobel Prize for 1938. Soon thereafter he came to the United States to live and was present at the 1939 conference at George Washington University at which Niels Bohr read a telegram from Lise Meitner, a noted German physicist, (who by that time was living in Stockholm) containing very exciting news. She told him that her former collaborators, Otto Hahn and Fritz Strassman, at Berlin University had found that a uranium nucleus hit by a neutron splits into two about equal parts, liberating vast amounts of energy. This announcement started a series of events which culminated, not too many years later, in nuclear bombs, nuclear power plants, etc., heralding the beginning of what one often calls the Atomic (Nuclear would be more correct) Age.

Fermi took the leadership in the top-secret laboratory at the University of Chicago, and on December 2, 1942 announced that on that afternoon the first chain reaction in uranium was achieved, thus initiating the first controlled release of nuclear energy by man.

Since this book is devoted to the progress of the understanding of the nature of things, and not to the prac-

Professor Paul Ehrenfest explaining a difficult point to his audience. (Photographed, probably, by Dr. S. Goudsmit)

tical applications, we will omit any discussion of fission chain reaction, and finish this chapter by describing an interesting experiment carried out by Fermi in the newly invented fission reactor. For the first time it became possible to measure the mean life of a neutron, which ultimately decays into a proton, an electron, and a neutrino. The gadget used in that experiment is known as the "Fermi bottle," although it was actually an evacuated spherical container somewhat resembling a Chianti bottle without a neck. As shown in Fig. 29, this sphere was placed inside an atomic pile and left

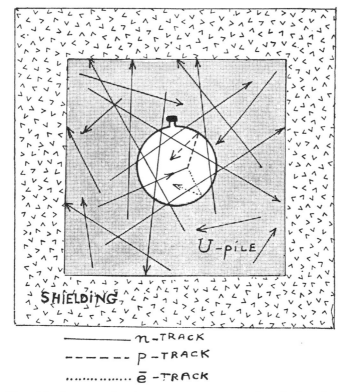

Fig. 29. Fermi's bottle in a uranium pile, designed to measure the neutron's mean life.

there for a considerably long period of time while the pile was in operation. The fission neutrons crisscrossing the pile in large quantities in most cases entered and left the "Fermi bottle," passing without much difficulty through its walls. Once in a while, however, a neutron passing through the "bottle" would break up into a proton and an electron for which the walls of the "bottle" were impenetrable. Thus ordinary hydrogen gas gradually accumulated inside the "bottle" at a rate which depended on the chance that a neutron would break up while passing through. Measuring the amount of hydrogen accumulated in the "bottle" during a given period of time, one could easily estimate the neutron's mean lifetime, which turned out to be about fourteen minutes.

To learn more about Fermi's activities along these lines the reader should turn to the book *Atoms in the Family,* written after his death by his wife, Laura.

CHAPTER VIII

H. YUKAWA AND MESONS

The great success of Fermi's theory of β-decay raised the question whether or not it could also be applied to explain the attractive forces holding nucleons together. It was known at the time that the forces between two nucleons—be they two neutrons, a neutron and a proton, or two protons—are identical except that in the last case one should add the ordinary Coulomb repulsion between the proton charges. Experiments have shown that in contrast to Coulomb forces, which decrease comparatively slowly with the distance (as $1/r^2$), nuclear forces are more similar to the cohesive forces of classical physics. Just as two pieces of Scotch tape do not exert any force on each other, no matter how close they are brought together, but stick tightly as soon as they come into contact, the forces between nucleons appear suddenly when they "touch each other," which occurs at the distance of about 10^{-13} cm. Once this happens, it takes about ten million electron volts of energy to separate them again. Similar forces acting between atoms are ascribed to the exchange of electrons between atomic shells as soon as they come into contact. The wave mechanical theory of these "exchange forces" was developed in 1927 by W. Heitler and F. London, who showed that the problem could be solved exactly in the simple case of two hydrogen atoms forming a diatomic molecule.† Heitler and London considered two cases: (a) a hydrogen molecule ion, H_2^+, consisting of two protons and one electron; (b) a neutral hydrogen molecule, H_2, formed by two protons and two electrons (Figs. 30a and 30b).

Schrödinger's wave equation for the electron's motion was solved exactly in both cases. The analytical result showed that there exists an equilibrium state of the lowest energy at certain distances R and R^1 between the two nuclei. The computed energy of these equilibrium states turned out to be in perfect agreement with the measured dissociation energies of H_2^+ and H_2 molecules. Thus, the notion of exchange forces between two identical atoms was firmly established in the field of quantum chemistry.

It was only natural to assume that attractive forces between two nucleons could be understood on a similar basis. When two nucleons were brought closely together, an electron accompanied by a neutrino would jump to and fro between them, thus producing an attractive exchange force. It was a very good idea, but (alas!) it did not work. When, in 1934, D. Ivanienko

† A diatomic molecule is one composed of two atoms.

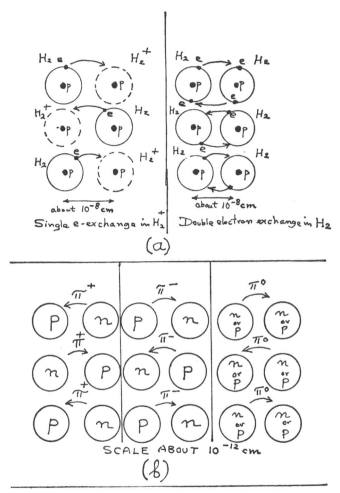

Fig. 30. Exchange forces. (a) Heitler and London's theory of the forces which hold together two protons in an ionized and a neutral hydrogen molecule. (b) Three different possibilities of explaining the nuclear forces by the exchange of pions ($\overset{+}{\pi}$, $\bar{\pi}$, $\overset{\circ}{\pi}$).

and I. Tamm computed the exchange force between two nucleons resulting from the Fermi interaction, they found that the expected binding energy was of the order of magnitude of 10^{-8} electron volts! No, this is not a misprint; one hundred millionth of one electron volt instead of ten million volts; just fifteen zeros too small! Apparently, Fermi's "weak" interaction could not be responsible for the strong binding of protons and neutrons within the atomic nucleus.

A year later (in 1935) a Japanese physicist, Hideki Yukawa, proposed a revolutionary idea to explain strong interactions between nucleons. If these interactions cannot be explained by the exchange forces arising from the to and fro jumping of electron-neutrino pairs, there must exist an entirely new, and as yet undetected, particle that does the jumping. To have the strength required by experimental evidence, that particle must be about two hundred times heavier than an electron (or about ten times lighter than a proton). Also, its interaction with nucleons, characterized by Yukawa's interaction constant y, must be about 10^{14} times greater than Fermi's interaction constant g responsible for β-transformations, thus being comparable to the ordinary Coulomb interactions between electric charges. This hypothetical particle was known under many aliases: Yukon, Japanese electron, heavy electron, mesotron, and finally meson. Two years later, after Yukawa's proposal, the particles with mass 207 times greater than that of electrons were found in cosmic rays by C. Anderson and S. Neddermeyer at the California Institute of Technology, and they seemed to give a brilliant confirmation to the Yukawa hypothesis. But then came a temporary setback. Experiments carried out by M. Converi, E. Pancini, and O. Piccioni proved without any doubt that, although the new particles had the mass of Yukawa's hypothetical mesons,

Typical Copenhagen Conference in 1930. First Row: Klein, Bohr, Heisenberg, Pauli, Gamow, Landau, Kramers.

their interaction with nucleons was 10^{12} times less than was needed to explain the nuclear forces. It was not until 1947 that the British physicist C. F. Powell, by sending photographic plates into the upper atmosphere, found that the mesons observed at sea level (207 electron masses) were actually the decay products of heavier mesons (mass 264 electron masses) which were formed by cosmic rays at the upper fringes of the terrestrial atmosphere. Thus there are two kinds of mesons: the heavy and the light ones. The former are now known as π-mesons, or simply pions, while the latter go under the name of μ-mesons, or muons for short. The pions show very strong interactions with nucleons and there is hardly any doubt that they *are* the particles initially visualized by Yukawa as being responsible for nuclear forces. However, no exact theory of these processes (comparable, for example, with Dirac's theory of anti-particles) has yet been developed.

CHAPTER IX

MEN AT WORK

The reader will have noticed that the chapters of this book have become shorter. This was not due to the author's growing fatigue but rather to the fact that, after the glorious developments in its first thirty years, Quantum Theory ran into serious difficulties, and its progress was considerably slowed down. The last "completely finished chapter" of this period was Dirac's unification of wave mechanics and Special Relativity, which resulted in the elegant theory of anti-particles. After they were found experimentally, anti-particles proved to behave exactly according to the theoretical predictions.

Fermi's theory of the processes involving the emissions and absorptions of electron-neutrino pairs becomes a little vague when applied to more complicated processes such as, for example, the decay of a muon into one electron and *two* neutrinos. Also, the numerical value of the Fermi constant g still cannot be derived from the values of the other fundamental constants of nature. (Similarly, the Rydberg constant R of old spectroscopy remained an empirical constant until Bohr published his theory of the hydrogen atom.)

Similar difficulties exist in the case of Yukawa's theory of strong interactions between elementary particles, and the numerical value of the constant y still remains unexplained. A huge number of new facts are continuously being discovered by experimental research, and a large number of empirical rules are formulated by introducing new notions such as "parity," "strangeness," etc. On the whole, the situation today resembles in many respects that existing in optics and in chemistry toward the end of the last century when the regularities in the spectral series and chemical valency properties of different elements were well known empirically but not at all understood theoretically. Things changed abruptly for the better when the Quantum Theory of atomic structure was developed and threw bright light on all the painfully collected empirical facts. In the opinion of the author, the present stalemate in the theory of elementary particles will be broken up—maybe next year, maybe in the year A.D. 2000—by an entirely new idea which will differ from the present way of thinking just as much as the present way differs from the classical one. We have no crystal ball for predicting the future developments of theoretical physics, but, as a substitute for it, one could use a discipline known as "dimensional analysis." Everybody knows that all physical measurements are based on *three basic units:*

> *Length* (stadia, miles, leagues, meters, etc.)
> *Time* (years, days, milliseconds, etc.)
> *Mass* (stones, pounds, drachms, grams, etc.)

Every physical quantity can be expressed through those three by the so-called "dimensional formulae." For example, *velocity* (*V*) is the length (or distance) covered per unit of time; *density* (*ρ*) is mass per unit of volume (i.e., length in the third power); *energy* (*E*) is the mass times the square of velocity; etc. One writes:

$$|V| = \frac{|L|}{|T|}; \ |\rho| = \frac{|M|}{|L|^3}; \ |E| = |M| \cdot \frac{|L|^2}{|T|}$$

where the vertical bars indicate that this is not a numerical relation but a relation between the physical nature of the quantities involved. It does not matter here which particular units one uses, and one can write:

$$|\$| \ = \ |\pounds| \ = \ |\text{mark}| \ = \ |\text{franc}| \ = \ |\text{ruble}| \ = \text{etc.}$$

or

$$|\text{yard}| \ = \ |\text{foot}| \ = \ |\text{meter}| \ = \ |\text{arshin}| \ = \ |\text{light-year}| \ = \text{etc.}$$

Length, time, and mass (or less correctly, weight) have been selected in classical physics on some kind of anthropomorphic basis, that is, on notions that we human beings encounter in everyday life. ("It is five miles away"; "I'll be back in an hour"; "Give me three pounds of ground round steak.") But the selection of these particular units is really not necessary, and *any three* complex units, be they the strength of the electric current (Amp.), the power of an engine (H.P.), or the

A typical Copenhagen Spring Conference (1932) at which the play Faust (a parody) was presented. In the first row: N. Bohr, P. A. M. Dirac, W. Heisenberg, P. Ehrenfest, M. Delbrück, Lisa Meitner, with many other great brains (guess them!) in the rows behind. (The author of this book missed the conference, being detained in the U.S.S.R.)

brightness of light (standard candle), can serve as basic units, provided they are independent of one another. However, in building a consistent theory of all physical phenomena, it is rational to select three fundamental units, each of which governs a vast area of physics, and to express all other units through them. Which units should be the members of this trio?

One of them doubtless should be the velocity of light in vacuum (c), which governs the entire field of electrodynamics and the Theory of Relativity. In fact, if one would assume that light propagates with infinite velocity $(c = \infty)$, Einstein's entire theory would reduce to the classical mechanics of Isaac Newton.

Another member of the Universal Trio is, of course, the quantum constant (h), which governs all atomic phenomena. If one assumes h to be equal to zero, one returns again to Newtonian mechanics. The great merit of Dirac is that he succeeded in uniting Relativity and Quantum Theories and in his equations c and h occupy equally honorable positions.

But what is the third universal constant needed to make the system of theoretical physics complete? One of the possible august candidates is, of course, Newton's gravitational constant. But closer considerations seem to indicate that this constant is not very suitable for collaborating with the other two to explain atomic and nuclear phenomena. Gravitational forces are very important in astronomy, explaining the motions of planets, stars, and galaxies. But in our human-size world gravitational attraction between material bodies is negligibly small, and one would be greatly surprised to see two apples placed a few inches apart on a table roll toward each other, driven by Newtonian attraction. Only extremely sensitive instruments permit us to measure gravitational attraction between two normal-size bodies.† In the atomic and nuclear world the forces of gravity are quite insignificant; some 10^{50} times smaller than electric and magnetic forces! It was once suggested by Dirac that Newton's "constant of gravity" is not really a constant but a variable which decreases in inverse proportion to the age of the Universe. And he may very well be right!

So what then? Which universal constant will occupy the third seat? We can just as well begin with ancient Greek philosophers, who first conceived the idea of the atom: the smallest amount of matter. In his book *The Analysis of Matter*,‡ Bertrand Russell writes:

> We might suppose, as Henri Poincaré once suggested, and as Pythagoras apparently believed, that space and time are granular, not continuous—i.e. the distance between two particles may always be an integral multiple of some unit, and so may the time between two events. Continuity in the percept is no evidence of continuity in the physical process.

In his book *The Physical Principles of the Quantum Theory*,§ Werner Heisenberg wrote:

> Although it is perhaps possible in principle to diminish space and time intervals without limit by refinement of measuring instruments, nevertheless for the principal discussion of the concepts of the wave theory it is advantageous to introduce finite values for values of space and time intervals involved in the measurements and only pass to the limit zero ($\Delta x \rightarrow 0$; $\Delta t \rightarrow 0$) for these intervals at the end of the calculations. It is possible that future developments of the Quantum Theory will show that the limit, zero, for such intervals is ab-

† See *Gravity* by G. Gamow, published in 1962 in this same series.
‡ New York: Dover Publications (1954), p. 235.
§ Chicago: University of Chicago Press (1930), p. 48.

straction without physical meaning; however, for the present, there seems to be no reason for imposing any limitations.

However, six years later Heisenberg changed his opinion about the thirteen words following "however" in his above-quoted sentence, and suggested that the "divergencies" occurring in various fields of the Quantum Theory might be eliminated through the introduction of an elementary length of the order of magnitude of 10^{-13} cm.

What does the word "divergency" mean? In mathematics this term pertains to "infinite series," that is, to endless sequences of numbers that are to be added together. For example, we can write:

$$1 + 2 + 3 + 4 + 5 + \text{(and so on to infinity)}.$$

Clearly, the result of the summation will be infinite. But what about:

$$1 + 1/2 + 1/3 + 1/4 + 1/5 + \text{(and so on to infinity)}?$$

It can be shown that this summation also becomes infinite, or *diverges,* as mathematicians say. On the other hand, the series:

$$1 + 1/1 + 1/2! + 1/3! + 1/4! + 1/5! + \text{(and so on to infinity)}$$

(where $n!$ means the product of all integers from 1 to n) *converges,* and is equal to 2.3026. . . . Similarly, the series

$$1 - 1/3! + 1/5! - 1/7! + \text{(and so on to infinity)}$$

converges to the value 0.

The results of calculations carried out in theoretical physics are often represented in the form of infinite series. If they converge, as they often do, we have a clear-cut answer and the definite numerical value for the physical quantity we are trying to calculate. But if the series diverges, the result makes no sense leading to the infinite value of the quantity under consideration. As an early example of such divergencies, let us consider a problem concerning the mass of an electron. If we visualize an electron as a tiny electrically charged sphere with the charge $e = 4.80 \times 10^{-10}$ electrostatic units and the radius r_0, classical electrostatics tells us that the energy of the electric field surrounding it is equal to $\frac{1}{2} \frac{e^2}{r_0}$. According to Einstein's Law of Equivalence of Mass and Energy, the mass of that field is $\frac{1}{2} \frac{e^2}{r_0 c^2}$. Since that mass should not exceed the observed mass m_0 of an electron ($= 0.9 \times 10^{-27}$ gm), it follows that:

$$\frac{e^2}{2r_0 c^2} \leqq m_0 \text{ or } r_0 \geqq \frac{e^2}{2m_0 c^2} = 2.82 \times 10^{-13} \text{cm}$$

However, if one assumes that the electron is a point charge ($r_0 = 0$), the mass of the electric field surrounding it becomes infinite! On the other hand, there are many good theoretical reasons for assuming that the electron is a point charge. Similar contradictions began to arise in larger numbers in the course of the further development of particle physics, and one was always arriving at divergent (infinite) results unless one cut the infinite mathematical series arising from straightforward calculations at a certain spot without any sufficient reason for doing so. Pauli humoristically called the work in this direction *"Die Abschneidungsphysik"* (Cut-off Physics).

It was characteristic that the cut-off had always to be done at distances of the order of 10^{-13} cm. When in later years the range of forces acting between the nucleons was measured experimentally with sufficient precision, it turned out to be 2.8×10^{-13} cm; that is, exactly the same as the so-called "classical radius of the electron," calculated theoretically on the assumption that its mass is entirely due to the electrostatic field surrounding it. It becomes more and more evident that there is a lower limit of distance, the elementary length λ anticipated by Pythagoras, Henri Poincaré, Bertrand Russell, Werner Heisenberg, and others, which is fundamental in physics. Just as no velocity can exceed that of light c, no mechanical action can be smaller than elementary action h, no distance can be smaller than elementary length λ, and no time interval can be shorter than elementary duration λ/c. When we know how to introduce λ (and λ/c) into the basic equations of theoretical physics, we will be able to state proudly: "Now at last we understand how matter and energy work!"

But, after the thirty fat years in the beginning of the present century, we are now dragging through the lean and infertile years, and looking for better luck in the years to come. In spite of all the efforts of the old-timers like Pauli, Heisenberg, and others, and those of the younger generation like Feynman, Schwinger, Gell-Mann, and others, theoretical physics has made very little progress during the last three decades, as compared with the three previous decades. The situation may best be characterized by a letter which Pauli wrote to the author of this book about an attempt he made with Heisenberg to explain the masses of various elementary particles which at that time were multiplying like rabbits. Here is an extract from the letter, the major part of which (omitted in this text) is devoted to the problems of fundamental biology:

UNIVERSITY OF CALIFORNIA

DEPARTMENT OF PHYSICS
BERKELEY 4, CALIFORNIA

March 1st, 1958

Dear Gamow,

Thanks for your letter of Feb. 24th. The stuff of Heisenberg and me is, as I believe, only so complicated for the reason, that we both have not yet understood it sufficiently. (There is no "paper" yet; but some other preprint, not yet determined for the publication, may be sent to physicists soon, to satisfie their curiosity and to prevent wild rumors.) In this sense you find enclosed my comment on Heisenberg's radio advertisement. (Please don't publish it in the press, but please do show it to other physicists and make it popular among them.)

Comment on Heisenberg's radio advertisement:

"This is to show the world, that I can paint like Titian."

Only technical details are missing.

V. Pauli

Fig. 31. Letter from Pauli.

Seven years have passed since this letter was written, hundreds of articles have been published on the problem of elementary particles, and still we are in darkness and uncertainty on the subject. Let us hope that in a decade or two, or, at least, just before the beginning of the twenty-first century, the present meager years of theoretical physics will come to an end in a burst of entirely new revolutionary ideas similar to those which heralded the beginning of the twentieth century.

FAUST

EINE HISTORIE

MANUSCRIPT AFTER: J. W. von Goethe

PRODUCED BY: The Task Force of the
"Institute for Theoretical Physics,"
Copenhagen

Motto:

Not to criticize . . .

N. Bohr

PROLOGUE

Between Heaven and Hell

PREFATORY REMARKS

The early decades of the present century witnessed the heady development of the Quantum Theory of the atom, and during that era the roads of theoreticians of all nationalities led, not to Rome, but to Copenhagen, the home city of Niels Bohr, who was the first to formulate the correct atomic model. It became customary at the end of each spring conference at Blegdamsvej† 15 (the then street address of Bohr's Institute of Theoretical Physics) to produce a stunt pertaining to recent developments in physics. The 1932 conference, which coincided with the tenth anniversary of Bohr's Institute, followed closely on the British physicist James Chadwick's discovery of a new particle *having the same mass as a proton but deprived of any electric charge.* Chadwick called it the *neutron,* the name which is now familiar to anybody interested in nuclear physics and in what is called, somewhat incorrectly, "atomic energy."

But there was some mixup in terminology. A few years earlier Wolfgang Pauli used the same name for a hypothetical particle which *had no mass and no charge* and was, in his opinion, necessary to explain the violation of the Law of Conservation of Energy observed experimentally in the processes of radioactive Beta-decay. "Pauli's Neutron" was the subject of hot discussions among the physicists, but these discussions were exclusively oral or carried on by private corre-

† Pronounced "Blí-dams-vī."

spondence, and the name was never "copyrighted" through appearance in any publication. Thus, when the discovery of Chadwick's heavy neutron was announced in his 1932 paper in *Nature,* the name of Pauli's weightless neutron had to be changed. Enrico Fermi proposed calling it the *neutrino,* which in Italian means a little neutron. In the following translation the name of Pauli's "neutron" in the original text is changed to the present name "neutrino," the existence of which had not at that time been demonstrated. Many physicists, especially Paul Ehrenfest, of Leiden, were very skeptical concerning Pauli's hypothetical neutrino, and it was only in 1955 that its existence was indisputably proved by the experiments of Fred Reines and Cloyd Cowan, of the Los Alamos Scientific Laboratory.

The pages that follow are the script of a play that was written and performed by several pupils of Bohr and staged at the spring meeting in 1932. (The author of this book was unable to participate in the play, the Soviet Russian Government having refused him a passport to attend the Copenhagen meeting.) The theme of this dramatic masterpiece has Pauli (*Mephistopheles*) trying to sell to the unbelieving Ehrenfest (*Faust*) the idea of the weightless neutrino (*Gretchen*).

The Blegdamsvej *Faust,* rendered into English by Barbara Gamow, is reproduced in this book as an important document pertaining to these turbulent years in the development of physics. The authors and the performers prefer to remain anonymous, except for J. W. von Goethe. Because of our failure to locate the original author or authors and the original artist, we are suggesting that the publisher deduct an appropriate moiety of the royalties which are to be paid, and hold this amount in escrow for a period of two or three years in the hope that the publication of the book may lead to the discovery of the author and/or artist. Failing that, the sum of money could be presented to the Institute's Library for the acquisition of new books.

Thanks are due to Professor Max Delbrück for his, kind help in the interpretation of certain parts of the play.

G.G.

In the German text of this physics Faust, *Goethe's rhythms and rhyme schemes (see comparable passages in the original* Faust*) were followed closely but not exactly. Certain poetic license has been taken here too, with the result that this English version falls somewhere between the other two. Unfortunately, some of the lines from the original* Faust*, which were used verbatim in the German physics version, could not be used here. Also, there were a few puns in the German language for which it was necessary to attempt English substitutions. Some of the passages in prose in the German physics version have here been converted into verse and appear as speeches by the Master of Ceremonies. This was with the idea of making this* Faust *more playable on the stage.*

There is an amusing confusion of identity throughout: Gretchen is at times referred to as Gretchen and at other times as The Neutrino; Faust sometimes as Faust and at other times as Ehrenfest. But it all adds to the fun, and nobody's the worse for it.

And by the way, if this should be played on the stage, it would seem a good idea for the different minor characters (be they Human or Physical Concepts) to wear signs indicating who they are: "Slater," "Darwin," "The Monopole," "The False Sign," etc. Otherwise, the audience will be hopelessly muddled.

B.G.

WHOM THE CHARACTERS REPRESENT

(*Note:* the Master of Ceremonies is played by Max Delbrück, German physicist)

ARCHANGEL EDDINGTON	A. Eddington, British astronomer
ARCHANGEL JEANS	J. Jeans, British astronomer
ARCHANGEL MILNE	E. A. Milne, British astronomer
MEPHISTOPHELES	W. Pauli, German physicist
THE LORD	Niels Bohr, Danish physicist
THE HEAVENLY HOSTS	"Extras"
FAUST	P. Ehrenfest, Dutch physicist
GRETCHEN	The Neutrino
OPPIE	R. Oppenheimer, American physicist
A TALL MAN	R. C. Tolman, American physicist
MILLIKAN-ARIEL	R. A. Millikan, American physicist
LANDAU (DAU)	L. Landau, Russian physicist
GAMOW	G. Gamow, Russian physicist
SLATER	J. C. Slater, American physicist

DIRAC   P. A. M. Dirac,
          British physicist

DARWIN   C. Darwin, British physicist

FOWLER   R. H. Fowler,
          British physicist

FOUR GRAY WOMEN   The Gauge Invariant, Fine
                    Structure Constant, Neg-
                    ative Energy, Singularity

FRIENDLY PHOTOGRAPHER   A friendly photographer

WAGNER   J. Chadwick,
          British physicist

MYSTICAL CHORUS   Everybody who can sing

The Blegdamsvej
Faust

✳

The THREE ARCHANGELS, THE LORD, THE HEAVENLY
HOSTS, *and* MEPHISTOPHELES

ARCHANGEL EDDINGTON
    As well we know, the Sun is fated
      In polytropic spheres to shine;[1]
    Its journey, long predestinated,
      Confirms *my theories* down the line.

    Hail to *Lemaître*'s promulgation[2]
      (Which none of us can understand)!
    As on the morning of Creation
      The brilliant Works are strange and grand.

ARCHANGEL JEANS
    And ever speeding and rotating,
      The double stars shine forth in flight,
    The Giants' brightness alternating
      With the eclipse's total night.

Ideal fluids, hot and spinning,
  By fission turn to pear-shaped forms.[3]
Mine are the theories that are winning!
  The atom cannot change the norms.

ARCHANGEL MILNE
  The storms break loose in competition
    (The *Monthly Notices* as well!)[4]
  And burn with violent ambition
    Important tidings to foretell.

  At heat of 10 to 7th power
    The gas degenerates in flame,
  Permitting us our shining hour
    Of freest flight in *Fermi*'s name.[5]

THE THREE
  This vision fills us with elation
    (Though none of us can understand).
  As on the Day of Publication
    The brilliant Works are strange and grand.

MEPHISTO
  (*springing forward*)

Since you, O *Lord,* yourself have now seen fit
To visit us and learn how each behaves,

And since it seems you favor me a bit,
Well—now you see me here

(*turning to the audience*)

                    among the slaves.
On Stars and Worlds I've nothing for the jury,
All that I know is how the folks complain.
To me the theory's full of sound and fury,
Yet here you are in ecstasy again,
Approving views that shatter like a bubble,
Sticking your nose in every kind of trouble.

THE LORD

But must you interrupt these revels
Just to complain, you Prince of Devils?
Does Modern Physics never strike you right?

MEPHISTO
No, Lord! I pity Physics only for its plight,
And in my doleful days it pains and sorely grieves
    me.
No wonder I complain—but who believes me?

THE LORD
You know this *Ehrenfest?* . . .

MEPHISTO
                    The Critic?[7]

(*A vision of the above appears*)

THE LORD

My knight!

MEPHISTO

Your knight, your slave and henchman. What's your
  bet?
You still will lose, I warn you, if you let
Me tempt this knight and lead him far astray.

THE LORD

Oh, this is really dreadful! Must I say . . .
Jah, muss Ich sagen. . . . There is an essential[8]
Failure of classic concepts—a morass.
One side remark—but keep it confidential—
Now what do you propose to do with *Mass?*

MEPHISTO

With Mass? Why, just forget it!

THE LORD
                          But . . . but this . . .
Is very ín-ter-est-ing. Yet to try it . . .

MEPHISTO
Oh, *Quatsch!* What rot you talk today! Be quiet!

THE LORD
But . . . but . . . but . . . but . . .

MEPHISTO
                          That's my hypothesis!

THE LORD
But *Pauli,* Pauli, Pauli, we practically agree.
There's no misunderstanding—that I guarantee.
Naturlich, Ich bin einig. We might throw *Mass* away
But *Charge* is something different—why Charge just
    has to stay!

MEPHISTO
What temperamental nonsense! Why *not* get rid of
Charge?

THE LORD
I   understand   completely,   but   *maa   jeg   spørge,*
friend,[9]

MEPHISTO
Shut up!

THE LORD
           But Pauli, surely you'll hear me to the end?
If Mass and Charge go packing, what have you, by
    and large?

MEPHISTO
Dear man, it's elementary! You ask me what remains?
Why bless me, The *Neutrino!* Wake up and use your
    brains!

(*Pause. Both pace to and fro*)

THE LORD

I say this not to criticize, but rather just to
learn. . . .[10]
But now I have to leave you. Farewell! I shall return!

(*He exits*)

MEPHISTO

From time to time it's pleasant to see the dear Old
Man,
I like to treat him nicely—as nicely as I can.
He's charming and he's lordly, a shame to treat him
foully—
And fancy!—he's so human he even speaks to Pauli!

(*He exits*)

FIRST PART

Faust's Study

FAUST

    I have—alas—learned Valence Chemistry,
    Theory of Groups, of the Electric Field,
    And Transformation Theory as revealed
    By Sophus Lie in eighteen-ninety-three.
    Yet here I stand, for all my lore,
    No wiser than I was before.
    M. A. I'm called, and Doctor. Up and down,
    Round and about, the pupils have been guided
    By this poor errin' Faust and witless clown;
    They break their heads on Physics, just as I did.
    But still I'm better than the cranks,
    The Big Shots, monkeys, mountebanks.
    All doubts assail me; so does *every* scruple;
    And Pauli as the Devil himself I fear.
    I grab the eraser, like a frantic pupil,
    Before the magic X-ings disappear,[11]
    For what is written down on black, in white,
    Is apt to be acceptable and right.
    Du Lieber Gott! I still could do some teaching.
    I have no *Guth* nor *Breit* here at my side,[12]
    But I could use their aptitude for preaching
    To spread the tested gospel *good* and *wide*.

Not even *Hund* nor *hound* could bear my lot,[18]
So I'm The Critic, sad and misbegot.

(MEPHISTO *bursts in
like thunder, dressed as
a traveling salesman*)

Why all the noise?

MEPHISTO

I'm at your service, Sir!

FAUST

What do you take me for? A customer?

MEPHISTO

You used to be receptive and urbane. . . .
These theories nowadays are wrong as rain;
Therefore I want to show you something higher,
For with it you can set the world on fire:
"The Dance of the Golden Calf"—kaleidoscopic—
The Radiation Theory is my topic.

(*Canon, sung by all*)

Born–Heisenberg
  Heisenberg–Pauli
    Pauli–Jordan
      Jordan–Wigner
        Wigner–Weisskopf
          Weisskopf–Born
            Born–Heisenberg[14] (*etc.*)
              (*etc.*)

MEPHISTO
These are my own,
Bone of my bone.
Listen how, with spunk and spice,
Precociously they give advice.
Here the width of lines diverges
In the wave-field's vasty length.

(*The* MASTER OF CEREMONIES *protests by gesture;*
MEPHISTO *repeats*)

Here the width of lines diverges
In the wave-field's loss of strength.

FAUST
Enough! You'll not seduce me. I am cured.
I'll never touch your reprints, rest assured.

MEPHISTO
I'm glad of that.

(*aside*)

      (His argument has pith.
The first old man that I can reason with!)

(*showing his wares*)

A Psi-Psi *Stern?*[15]

FAUST

No sale!

MEPHISTO

A Psi-Psi *Gerlach?*

FAUST

No sale!

MEPHISTO

Electrodynamics?

FAUST

No sale!

MEPHISTO

By *Heisenberg-Pauli?*

FAUST

No sale!

MEPHISTO

With infinite self-energy?

FAUST

No sale!

MEPHISTO

Electrodynamics?

FAUST

No sale!

MEPHISTO

> By *Dirac?*

FAUST

> No sale!

MEPHISTO

> With infinite self-energy?

FAUST

> The same old story!

MEPHISTO

> So I must show you something that's unique!

FAUST

> You'll not seduce me, softly though you speak.
> If ever to a theory I should say:
> "You are so beautiful!" and "Stay! Oh, stay!"
> Then you may chain me up and say goodbye—
> Then I'll be glad to crawl away and die.

MEPHISTO

> Beware alone of Reason and of Science,
> Man's highest powers, unholy in alliance.
> You let yourself, through dazzling witchcraft, yield
> To all temptations of the Quantum field.
> Listen! As now the obstacles abate,
> You'll know the fair *Neutrino* for your fate!

GRETCHEN

(comes in and sings to FAUST. *Melody: "Gretchen at the Spinning Wheel" by Schubert)*

My Mass is zero,
    My Charge is the same.
You are my hero,
    Neutrino's my name.

I am your fate,
    And I'm your key.
Closed is the gate
    For lack of me.

Beta-rays throng[16]
    With me to pair.
The N-spin's wrong[17]
    If *I'm* not there.

My Mass is zero,
    My Charge is the same.
You are my hero,
    Neutrino's my name.

My psyche turns
    To you, my own.
My poor heart yearns
    For you alone.

My lovesick soul
  Is yours to win.
I can't control
  My trembling spin.

My Mass is zero,
  My Charge is the same.
You are my hero,
  Neutrino's my name.

(Exeunt omnes)

MRS. ANN ARBOR'S SPEAKEASY[18]
(*otherwise known as Auerbach Keller*)

(*American physicists sitting sadly at the Bar*)

MEPHISTO

(*springing forward behind the bar*)

Can no one laugh? Will no one drink?
I'll teach you Physics in a wink. . . .

(*he winks exaggeratedly and knowingly at the physicists*)

Shame on you, sitting in a daze
When as a rule you're all ablaze!

OPPIE

(*swallowing*—Njum! Njum!—*before speaking*)

Your fault! You've brought no single word of cheer—
No news, no X-ings. Bah!

MEPHISTO

(*producing* GRETCHEN)

But both are here!

(*Lively applause and general tumult*)

A TALL MAN
A shapely and appealing Signorina. . . .

(*to* MEPHISTO)

But tell me, have you been in Pasadena?

MEPHISTO
With *Einstein*, yes. He greets you in your harbor,
This *wunder*-bar of Mrs. Annie Arbor.

A TALL MAN
Einstein! His curves! His fields! His whole arena!

MEPHISTO

(*sings*)

A *Monarch* cherished dearly
  A *Flea*, just as a son,[19]
And quite as much—or nearly—
  As Gra-vee-táy-shee-un.

The Monarch summoned *Mayer*,[20]
  Said Mayer: "To be sure!
I'll make him tensors, Sire,[21]
  With junker curvature."

Attired as a dandy,
  The Flea was then displayed.
Folks ate him up like candy
  So sweetly was he made.

The Flea grew up, and later
  His *Son* was born. The son[22]
Kept challenging his pater
  But never got to run.

$$\underline{\Gamma_{st}^{i}} = \Gamma_{st}^{i} + \Gamma_{st,r}^{i}\ \xi^{\ r}$$

$$\int \mathfrak{M}_i\ \xi^{\ i}\ d\tau$$

$$\mathfrak{g}^{\ is}_{\ \ s} = 0$$

Half-naked, fleas came pouring
   From Berlin's joy and pride,
Named by the unadoring:
   "Field Theories—Unified."

Now, Physicists, take warning,
   Observe this sober test. . . .
When new fleas are a-borning
   Make sure they're fully dressed!

ALL

Drunk though we are, we feel as fine
As—hic!—five hundred female swine!

FAUST

(*known to be opposed to alcohol, steps forward and sings*)

(*to* MEPHISTO)

Do you expect me to get well
In all this chaos, din and hell?

(*to* GRETCHEN)

You Skeleton, you Monster, here I stand,
But do you recognize your lord and master?
What holds me back? See here, I take your hand
And *shatter* you!

GRETCHEN

Faust, Faust, I fear disaster!

(*Exeunt omnes*)

SECOND PART

A Charming Region

(FAUST *sleeps, on a bed of roses. A plum tree grows,
to the right. A terrific din announces the approach of
the* MILLIKAN-ARIEL)

MILLIKAN-ARIEL

(*from above*)

Hear, oh hear the words of rubes
(Wilson Chambers, Counting Tubes)![23]
Thundering, for the spirit's ear,
Cosmic Rays will now appear!
The protons are creaking and chattering,
Electrons are rolling and clattering.
Light comes rushing—whither? whence?
Heisenberg is really grumpy;[24]
Rossi, Hoffmann—both are jumpy.[25]
All this nonsense makes no sense!

FAUST

(*awakening*)

Sweet rosy field—what soil am I caressing?
And why familiar? *Rosenfeld,* they say,[26]
To the green*gauge invariant* gives a blessing.[27]
This is his plum.

(MASTER OF CEREMONIES *appears*)

(*to the* M.C.)

What's going on today?

M.C.
Walpurgis Nights: the *Classical Poetical,*
And afterwards, the *Quantum Theoretical.*

FAUST
Excellent! I quite agree!

THE CLASSICAL WALPURGIS NIGHT

M.C.

(*makes a gesture of presentation*)

The Classical—a potpourri!

FAUST

(*He leans forward, expecting. A long pause indicates that nothing is happening*)

But nothing's happening!

M.C.

Just wait and see!

FAUST

(*He waits. Another long pause and again nothing happens*)

See here now, Delbrück! . . .

M.C.

Faust, you must expect
That with the Classical there's no effect
Upon the audience.

(DIRAC *enters*)

DIRAC

Correct! Correct!

FAUST

Why not skip this, and go to the Q.-T.?

M.C.

    If we do that, I fail as an M.C.,
    For first the Classical must duly close.

FAUST

    I have two different time-scales to propose
    For these Walpurgis Nights. As I've avowed,
    The First should go to limbo.

DIRAC

                                                 Not allowed!

FAUST

    I then propose the Classical be moved
    Much farther back in time and place.

M.C.

                                                 Approved!

THE QUANTUM THEORETICAL WALPURGIS NIGHT

(At one side of the stage, to the back, THE LORD *and* LANDAU[28] *appear, the latter bound and gagged)*

THE LORD

Keep quiet, *Dau!* . . . Now, in effect,
The only theory that's correct,
Or to whose lure I can succumb
Is

LANDAU

Um! Um-um! Um-um! Um-um!

THE LORD

Don't interrupt this colloquy!
I'll do the talking. Dau, you see,
The only proper rule of thumb
Is

LANDAU

Um! Um-um! Um-um! Um-um!

(*At the other side of the stage, to the back, appears the face of* GAMOW, *through bars*)

GAMOW

> I cannot go to Blegdamsvej
> (Potential barrier too high!).
> This "conversation" is the hoak—
> The Lord, he really make the joke.
> Bounded and gaggled, mouse to toe,
> Dau can't say "*Nyet!*" nor "*Horosho!*"

M.C.

(center stage)

> Be careful! *Achtung!* Watch it! These *Holes* of P.
>   Dirac[29]
> Can trip you in a second and flip you on your back!

(He puts up a "Warning!" sign)

THE MONOPOLE

(steps forward and sings)

> Two Monopoles worshiped each other,[30]
> And all of their sentiments clicked.
> Still, neither could get to his brother,
> Dirac was so fearfully strict!

(to the M.C.)

> But tell me—(Watch it! There's a Hole!)
> Where is my darling *Antipole?*

M.C.

(*aside*)

(A *Hole!* My foot! More like a crater!)

(*to the* MONOPOLE)

Now just a minute—Here comes *Slater.*

(SLATER *steps forward with a bloody lance and* THE
GROUP DRAGON)[31]

M.C.

(*observing the characters running about on the
stage*)

Why do they run? Why does he roll?
Who stabbed him with the bloody pole?
  Group Dragon, by this mortal blow
  We laid you low!

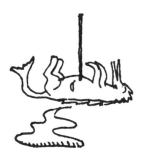

Scaly with indices is he
Who died of Anti-symmetry.

Reduced to nothing, there he lies
Stripped of his status and disguise.
Group Dragon, by this mortal blow
We laid you low!

(*The* FALSE SIGN *steps forward*)

FALSE SIGN[32]

All the theories expire or bring disappointment.
The Sign is forever the fly in your ointment.
The reckoning's splendid and everything's fine—
Nonetheless, at the zenith, in squeezes The Sign!

(DIRAC *and* DARWIN *are brought forward*)

M.C.

Now here's the revered *One-dimensional Case;*
His name is *Dirac*—you remember his face.
Three-dimensional Darwin is following next.

(*The* FALSE SIGN *prances around* DIRAC *and pulls
him to the side. But he has no access to* DARWIN)

Observe the False Sign; he's annoyed and perplexed.
This injures his pride. He can handle Dirac,
But Darwin's a nut that he can't seem to crack,
For Darwin so far is like pie in the sky—
He's only a glint in a physicist's eye.

(M.C. *holds up a card that reads*)

> THE   EXCHANGE   RELATION[38]
> $PQ - QP = h/2\pi i$

Watch this! Darwin's turned himself into a *P*,

(FOWLER *arrives on the scene*)

And *Fowler*—he's *Q*—has arrived. As you'll see,
They explain the *Relation* described on the card
By leapfrogging madly all over the yard.

(*At each exchange flashes the sign* "h/2πi." *With this goes a song*):

Thus exchanged are *P* and *Q*
   Time and time anew,
   Time and time anew.
Still there ever hovers by:
   h/2πi, h/2πi!

They can never rest in peace
　　Till they're gone as geese,
　　Till they're gone as geese.
Still there ever hovers by:
　　$h/2\pi i$, $h/2\pi i!$

Attention! Attention! Their form is now altered

(*P and Q now suffer the painful metamorphosis into* DONKEY-ELECTRONS *and fall into one of Dirac's Holes*)

To *Donkey-electrons.* Observe that they've fal-
　　tered[34]
And fallen, through carelessness (clumsy old
　　chaps!),
Into one of those Holes that are planted as traps.

(*The* SPIN OF THE PHOTON, *dressed in Indian guise, slithers across the stage, accompanied by fugitive music*)

Attention again! Here's *The Spin of the Photon*[35]
With some kind of Indian *sari* and coat on.
(It's clear that no modest, respectable *Boson*[36]
Would traverse the platform without any clo'es on!)

(DIRAC *comes forward, followed by* FOUR GRAY WOMEN)

THE FIRST
The *Gauge Invariant* is my name.

THE SECOND
I'm of *Fine Structure Constant* fame.[87]

THE THIRD
Negative Energy—that's me.[88]

THE FOURTH

(*to* THE THIRD)

Just watch your grammar, Number Three!

(*to the others*)

Sisters, into the reckoning
You cannot and you may not spring.
But in the end there I shall be,
For I am *Singularity!*[89]

(THE FOUR *stand to the side of the stage, to mingle in again later*)

FAUST
Four I saw come, one I saw go;
And what they tried to say I do not know.
The air is now so full of shades and spooks
That we had best hang on to our perukes.

DIRAC
A strange bird croaks. It croaks of what? Bad luck!
Our theories, gentlemen, have run amuck.
To 1926 we must return;[40]
Our work since then is only fit to burn.

FAUST
Then nothing should originate today?

DIRAC

(*to the* FOURTH GRAY WOMAN)

You, Singularity, just *go away!*

THE FOURTH
My place is here—and, if you please, *don't shout!*

DIRAC
Wench, through my magic I will get you out!

THE FOURTH
Am I not in Eigen fields?
Does Radiation not contain me?
My form to change forever yields,
My power is such that none can chain me.
Yet on the track, as on the waves,
I stand among the frightened slaves,
Always found, though never sought,
Cursed before she's even caught.

DIRAC

I don't see your point!

(*He exits, chased by* SINGULARITY)

M.C.

(*to* DIRAC'S *back*)

You'll see it soon—
That woman's going to chase you to the Moon!

(*to the audience*)

Unless, of course, his long legs save the day.
Three guesses! Will he make his getaway?

(MEPHISTO *appears.* Somebody knocks at the door.
A FRIENDLY PHOTOGRAPHER *looks questioningly inside.*)

MEPHISTO

Come on, come on! Come in, come in!
You baggy-trousered lout, you,
With plate and film and click and din!

(*pointing to* FAUST)

He shrivels up without you.

FAUST

(*Highly excited, he takes a pose for the press photographer*)

To this fair moment let me say:
"You are so beautiful—Oh, stay!"
A trace of me will linger 'mongst the Great,
Within the annals of The Fourth Estate.
Anticipating fortune so benign,
I now enjoy the moment that is mine!

(*He dies, and his body is carried out by the Press*)

MEPHISTO

No pleasure was enough; no luck appeased him.
The changing forms he wooed have never pleased him.
The poor man clung to those who would evade him.
All's over now. How did his knowledge aid him?

M.C.

(*to the photographer's camera*)

Out, Light overpowering!—
Magnesium-devouring,
Thundercloud-showering,
Ego-deflowering,
   Stinking One,
   Blinking One,
Vex us no more!

FINALE

Apotheosis of the True Neutron

WAGNER[41]

(appears, as the personification of the ideal experimentalist, balancing a black ball on his finger, and says, with pride)

The *Neutron* has come to be.
Loaded with Mass is he.
Of Charge, forever free.
Pauli, do you agree?

MEPHISTO
      That which experiment has found—
      Though theory had no part in—
      Is always reckoned more than sound
      To put your mind and heart in.
      Good luck, you heavyweight Ersatz—[42]
      We welcome you with pleasure!
      But passion ever spins our plots,
      And Gretchen is my treasure!

MYSTICAL CHORUS
      Now a reality,
      Once but a vision.
      What classicality,
      Grace and precision!
      Hailed with cordiality,
      Honored in song,
      Eternal Neutrality
      Pulls us along!

Solvay International Institute of Physics, Sixth Council of Physics, Brussels,
October 20–26, 1930.

First Row: Th. De Donder, P. Zeeman, P. Weiss, A. Sommerfeld, Mme. Curie,
P. Langevin, A. Einstein, O. Richardson, B. Cabrera, N. Bohr, W. J. De Haas.
Second Row: E. Herzen, E. Henriot, J. Verschaffelt, Manneback, A. Cotton, J. Errera,
O. Stern, A. Piccard, W. Gerlach, C. Darwin, P. A. M. Dirac, H. Bauer, P. Kapitza,
L. Brillouin, H. A. Kramers, P. Debye, W. Pauli, J. Dorfman, J. H. Van Vleck, E. Fermi,
W. Heisenberg. (Photographed by Benjamin Couprie)

Solvay International Institute of Physics, Seventh Council of Physics, Brussels, October 22–29, 1933.
First Row: E. Schrödinger, Mme. I. Joliot, N. Bohr, A. Joffé, Mme. Curie, O. W. Richardson, P. Langevin, Lord Rutherford, Th. De Donder, M. de Broglie, L. de Broglie, Mlle. L. Meitner, J. Chadwick.
Second Row: E. Henriot, F. Perain, F. Joliot, W. Heisenberg, H. A. Kramers, E. Stahel, E. Fermi, E. T. S. Walton, P. A. M. Dirac, P. Debye, N. F. Mott, B. Cabrera, G. Gamow, W. Bothe, P. Blackett, M. S. Rosenblum, J. Errera, Ed. Bauer, W. Pauli, J. E. Verschaffelt, M. Cosyns (in back), E. Herzen, J. D. Cockroft, C. D. Ellis, R. Peierls, Aug. Piccard, E. O. Lawrence, L. Rosenfeld. (Photographed by Benjamin Couprie)

NOTES ON THE TEXT

PROLOGUE

1. Polytropic spheres are mathematical models of hot gaseous spheres representing stars.
2. Abbé Georges Lemaître, a Belgian astronomer who originated the Theory of the Expanding Universe.
3. . . . pear-shaped forms. Jeans' theory of the origin of double stars.
4. *Monthly Notices* of the Royal Astronomical Society, in which most British papers on theoretical astrophysics are published.
5. Fermi's degenerated electron gas forms the interior of certain classes of stars (see Chapter VII).
6. A symbol characterizing the quantum theoretical uncertainty relation $\Delta q \cdot \Delta p$ which will be encountered later in the text.
7. The Critic. Professor Ehrenfest had a critical attitude toward many theoretical ideas, in particular toward Pauli's hypothesis of the neutrino.
8. *Jah, muss Ich sagen.* . . . The accepted language in Bohr's Institute was to a large extent German, because of the many visitors from Central Europe. Bohr spoke it perfectly but often with Danishisms. One of his typical expressions was *"muss Ich sagen"* ("must I say") which should have been, in correct German, *"darf Ich sagen"* ("may I say"). The reason was that in Danish *"darf"* is *"maa,"* which is closer to the German *"muss"* or the English "must."
9. *"maa jeg spørge"* means "may I ask" in Danish.
10. ". . . not to criticize." Another of Bohr's typical ex-

pressions which he always used when he did not agree with someone's statement.

FIRST PART

11. X-ings. *("Ixerei"* in German) is a word invented by Einstein and was often used about papers which contained too much complicated mathematics, ("X" is the unknown in school algebra) but little physical content.
12. E. Guth (in English his name means "good") and G. Breit (in English, "wide").
13. The German physicist, F. Hund (in English, "dog") whose name was often used in the expression "working like a dog."
14. German physicists working on the Quantum Theory of Radiation.
15. Psi-Psi Stern (in English, "Psi-Psi Star—$\psi\psi^*$) is an important quantity in quantum physics. Here referring to Otto Stern and W. Gerlach, well-known experimentalists.
16. Beta-rays. According to Pauli's hypothesis, the neutrino is a particle which always accompanies the emission of a Beta-ray from the nucleus.
17. N-spin. According to the views of those days, the spin (axial rotation) of the nitrogen nucleus could not be explained without considering the spin of the hypothetical neutrino.
18. Referring to the University of Michigan at Ann Arbor, Michigan.
19. The Flea of the Monarch (Einstein) is the General Theory of Relativity.
20. Walter Mayer—pronounced "Myer" (rhymes with "Sire"!)—mathematician who assisted Einstein in the development of his theories.
21. Tensors are mathematical symbols used in the Theory of Curved Spaces.
22. The Son is the Unified Field Theory on which Einstein

worked for the last three decades of his life, but
without much success.

SECOND PART

23. Wilson Chambers, etc. Physical apparatuses used in the
    study of cosmic rays.
24. W. Heisenberg (see Chapter VI), who was at that
    period interested in the Theory of Cosmic Rays.
25. Bruno Rossi and G. Hoffmann, experimentalists study-
    ing cosmic rays.
26. Leon Rosenfeld, a Belgian theoretical physicist.
27. "Gauge invariant," a complicated notion in theoretical
    physics. In German it is called *"Eiche invariant."* By
    coincidence the German word *"Eiche"* also means
    "oak."
28. Landau. See *Niels Bohr and the Development of
    Physics,* ed. by W. Pauli, New York, McGraw-Hill
    Book Co., Inc., 1955, page 70.
29. Dirac's Holes (see Chapter VII for explanation).
30. Monopole (see Chapter VII for explanation).
31. The Group Theory, a complicated branch of mathe-
    matics, used in the Quantum Theory.
32. The False Sign refers to the putting of a $+$ instead of
    a $-$, or vice versa, a mistake often made by absent-
    mindedness in mathematical calculations and lead-
    ing, of course, to wrong results.
33. The Exchange Relation. The basic postulate in the
    Heisenberg quantum mechanics.
34. Donkey-electrons. A jocular expression for an electron
    with negative mass (see Chapter VII).
35. A photon, or light quantum, can be considered as a
    rotating package of energy.
36. Re bosons (see Chapter IV).
37. Fine Structure Constant. The number 137, which is
    important in the theory of the atom.
38. Negative Energy. One of the mathematical difficulties
    appearing in the Quantum Theory.

39. Singularity. Another mathematical difficulty appearing in the Quantum Theory.
40. In the year 1926 wave mechanics was discovered.

FINALE

41. Wagner. James Chadwick, a British physicist who discovered the neutron (heavy neutral particle) in the year this play was presented.
42. Ersatz. The neutron, with its large mass, cannot be considered as a substitute ("Ersatz") for the weightless neutrino.

INDEX

Numbers appearing in italics refer to numbers of the figures in the text.

A CATALOG OF SELECTED
DOVER BOOKS
IN ALL FIELDS OF INTEREST

A CATALOG OF SELECTED DOVER
BOOKS IN ALL FIELDS OF INTEREST

DRAWINGS OF REMBRANDT, edited by Seymour Slive. Updated Lippmann, Hofstede de Groot edition, with definitive scholarly apparatus. All portraits, biblical sketches, landscapes, nudes. Oriental figures, classical studies, together with selection of work by followers. 550 illustrations. Total of 630pp. 9⅛ × 12¼.
21485-0, 21486-9 Pa., Two-vol. set $25.00

GHOST AND HORROR STORIES OF AMBROSE BIERCE, Ambrose Bierce. 24 tales vividly imagined, strangely prophetic, and decades ahead of their time in technical skill: "The Damned Thing," "An Inhabitant of Carcosa," "The Eyes of the Panther," "Moxon's Master," and 20 more. 199pp. 5⅜ × 8½. 20767-6 Pa. $3.95

ETHICAL WRITINGS OF MAIMONIDES, Maimonides. Most significant ethical works of great medieval sage, newly translated for utmost precision, readability. Laws Concerning Character Traits, Eight Chapters, more. 192pp. 5⅜ × 8½.
24522-5 Pa. $4.50

THE EXPLORATION OF THE COLORADO RIVER AND ITS CANYONS, J. W. Powell. Full text of Powell's 1,000-mile expedition down the fabled Colorado in 1869. Superb account of terrain, geology, vegetation, Indians, famine, mutiny, treacherous rapids, mighty canyons, during exploration of last unknown part of continental U.S. 400pp. 5⅜ × 8½. 20094-9 Pa. $6.95

HISTORY OF PHILOSOPHY, Julián Marías. Clearest one-volume history on the market. Every major philosopher and dozens of others, to Existentialism and later. 505pp. 5⅜ × 8½. 21739-6 Pa. $8.50

ALL ABOUT LIGHTNING, Martin A. Uman. Highly readable non-technical survey of nature and causes of lightning, thunderstorms, ball lightning, St. Elmo's Fire, much more. Illustrated. 192pp. 5⅜ × 8½. 25237-X Pa. $5.95

SAILING ALONE AROUND THE WORLD, Captain Joshua Slocum. First man to sail around the world, alone, in small boat. One of great feats of seamanship told in delightful manner. 67 illustrations. 294pp. 5⅜ × 8½. 20326-3 Pa. $4.95

LETTERS AND NOTES ON THE MANNERS, CUSTOMS AND CONDITIONS OF THE NORTH AMERICAN INDIANS, George Catlin. Classic account of life among Plains Indians: ceremonies, hunt, warfare, etc. 312 plates. 572pp. of text. 6⅛ × 9¼. 22118-0, 22119-9 Pa. Two-vol. set $15.90

ALASKA: The Harriman Expedition, 1899, John Burroughs, John Muir, et al. Informative, engrossing accounts of two-month, 9,000-mile expedition. Native peoples, wildlife, forests, geography, salmon industry, glaciers, more. Profusely illustrated. 240 black-and-white line drawings. 124 black-and-white photographs. 3 maps. Index. 576pp. 5⅜ × 8½. 25109-8 Pa. $11.95

THE BOOK OF BEASTS: Being a Translation from a Latin Bestiary of the Twelfth Century, T. H. White. Wonderful catalog real and fanciful beasts: manticore, griffin, phoenix, amphivius, jaculus, many more. White's witty erudite commentary on scientific, historical aspects. Fascinating glimpse of medieval mind. Illustrated. 296pp. 5⅜ × 8¼. (Available in U.S. only) 24609-4 Pa. $5.95

FRANK LLOYD WRIGHT: ARCHITECTURE AND NATURE With 160 Illustrations, Donald Hoffmann. Profusely illustrated study of influence of nature—especially prairie—on Wright's designs for Fallingwater, Robie House, Guggenheim Museum, other masterpieces. 96pp. 9¼ × 10¾. 25098-9 Pa. $7.95

FRANK LLOYD WRIGHT'S FALLINGWATER, Donald Hoffmann. Wright's famous waterfall house: planning and construction of organic idea. History of site, owners, Wright's personal involvement. Photographs of various stages of building. Preface by Edgar Kaufmann, Jr. 100 illustrations. 112pp. 9¼ × 10.
23671-4 Pa. $7.95

YEARS WITH FRANK LLOYD WRIGHT: Apprentice to Genius, Edgar Tafel. Insightful memoir by a former apprentice presents a revealing portrait of Wright the man, the inspired teacher, the greatest American architect. 372 black-and-white illustrations. Preface. Index. vi + 228pp. 8¼ × 11. 24801-1 Pa. $9.95

THE STORY OF KING ARTHUR AND HIS KNIGHTS, Howard Pyle. Enchanting version of King Arthur fable has delighted generations with imaginative narratives of exciting adventures and unforgettable illustrations by the author. 41 illustrations. xviii + 313pp. 6⅛ × 9¼. 21445-1 Pa. $6.50

THE GODS OF THE EGYPTIANS, E. A. Wallis Budge. Thorough coverage of numerous gods of ancient Egypt by foremost Egyptologist. Information on evolution of cults, rites and gods; the cult of Osiris; the Book of the Dead and its rites; the sacred animals and birds; Heaven and Hell; and more. 956pp. 6⅛ × 9¼.
22055-9, 22056-7 Pa., Two-vol. set $20.00

A THEOLOGICO-POLITICAL TREATISE, Benedict Spinoza. Also contains unfinished *Political Treatise*. Great classic on religious liberty, theory of government on common consent. R. Elwes translation. Total of 421pp. 5⅜ × 8½.
20249-6 Pa. $6.95

INCIDENTS OF TRAVEL IN CENTRAL AMERICA, CHIAPAS, AND YUCATAN, John L. Stephens. Almost single-handed discovery of Maya culture; exploration of ruined cities, monuments, temples; customs of Indians. 115 drawings. 892pp. 5⅜ × 8½. 22404-X, 22405-8 Pa., Two-vol. set $15.90

LOS CAPRICHOS, Francisco Goya. 80 plates of wild, grotesque monsters and caricatures. Prado manuscript included. 183pp. 6⅝ × 9⅝. 22384-1 Pa. $4.95

AUTOBIOGRAPHY: The Story of My Experiments with Truth, Mohandas K. Gandhi. Not hagiography, but Gandhi in his own words. Boyhood, legal studies, purification, the growth of the Satyagraha (nonviolent protest) movement. Critical, inspiring work of the man who freed India. 480pp. 5⅜ × 8½. (Available in U.S. only)
24593-4 Pa. $6.95

CATALOG OF DOVER BOOKS

ILLUSTRATED DICTIONARY OF HISTORIC ARCHITECTURE, edited by Cyril M. Harris. Extraordinary compendium of clear, concise definitions for over 5,000 important architectural terms complemented by over 2,000 line drawings. Covers full spectrum of architecture from ancient ruins to 20th-century Modernism. Preface. 592pp. 7½ × 9⅝. 24444-X Pa. $14.95

THE NIGHT BEFORE CHRISTMAS, Clement Moore. Full text, and woodcuts from original 1848 book. Also critical, historical material. 19 illustrations. 40pp. 4⅝ × 6. 22797-9 Pa. $2.25

THE LESSON OF JAPANESE ARCHITECTURE: 165 Photographs, Jiro Harada. Memorable gallery of 165 photographs taken in the 1930's of exquisite Japanese homes of the well-to-do and historic buildings. 13 line diagrams. 192pp. 8⅜ × 11¼. 24778-3 Pa. $8.95

THE AUTOBIOGRAPHY OF CHARLES DARWIN AND SELECTED LETTERS, edited by Francis Darwin. The fascinating life of eccentric genius composed of an intimate memoir by Darwin (intended for his children); commentary by his son, Francis; hundreds of fragments from notebooks, journals, papers; and letters to and from Lyell, Hooker, Huxley, Wallace and Henslow. xi + 365pp. 5⅜ × 8. 20479-0 Pa. $6.95

WONDERS OF THE SKY: Observing Rainbows, Comets, Eclipses, the Stars and Other Phenomena, Fred Schaaf. Charming, easy-to-read poetic guide to all manner of celestial events visible to the naked eye. Mock suns, glories, Belt of Venus, more. Illustrated. 299pp. 5¼ × 8¼. 24402-4 Pa. $7.95

BURNHAM'S CELESTIAL HANDBOOK, Robert Burnham, Jr. Thorough guide to the stars beyond our solar system. Exhaustive treatment. Alphabetical by constellation: Andromeda to Cetus in Vol. 1; Chamaeleon to Orion in Vol. 2; and Pavo to Vulpecula in Vol. 3. Hundreds of illustrations. Index in Vol. 3. 2,000pp. 6⅛ × 9¼. 23567-X, 23568-8, 23673-0 Pa., Three-vol. set $38.85

STAR NAMES: Their Lore and Meaning, Richard Hinckley Allen. Fascinating history of names various cultures have given to constellations and literary and folkloristic uses that have been made of stars. Indexes to subjects. Arabic and Greek names. Biblical references. Bibliography. 563pp. 5⅜ × 8½. 21079-0 Pa. $7.95

THIRTY YEARS THAT SHOOK PHYSICS: The Story of Quantum Theory, George Gamow. Lucid, accessible introduction to influential theory of energy and matter. Careful explanations of Dirac's anti-particles, Bohr's model of the atom, much more. 12 plates. Numerous drawings. 240pp. 5⅜ × 8½. 24895-X Pa. $4.95

CHINESE DOMESTIC FURNITURE IN PHOTOGRAPHS AND MEASURED DRAWINGS, Gustav Ecke. A rare volume, now affordably priced for antique collectors, furniture buffs and art historians. Detailed review of styles ranging from early Shang to late Ming. Unabridged republication. 161 black-and-white drawings, photos. Total of 224pp. 8⅜ × 11¼. (Available in U.S. only) 25171-3 Pa. $12.95

VINCENT VAN GOGH: A Biography, Julius Meier-Graefe. Dynamic, penetrating study of artist's life, relationship with brother, Theo, painting techniques, travels, more. Readable, engrossing. 160pp. 5⅜ × 8½. (Available in U.S. only) 25253-1 Pa. $3.95

HOW TO WRITE, Gertrude Stein. Gertrude Stein claimed anyone could understand her unconventional writing—here are clues to help. Fascinating improvisations, language experiments, explanations illuminate Stein's craft and the art of writing. Total of 414pp. 4⅝ × 6⅝. 23144-5 Pa. $5.95

ADVENTURES AT SEA IN THE GREAT AGE OF SAIL: Five Firsthand Narratives, edited by Elliot Snow. Rare true accounts of exploration, whaling, shipwreck, fierce natives, trade, shipboard life, more. 33 illustrations. Introduction. 353pp. 5⅜ × 8½. 25177-2 Pa. $7.95

THE HERBAL OR GENERAL HISTORY OF PLANTS, John Gerard. Classic descriptions of about 2,850 plants—with over 2,700 illustrations—includes Latin and English names, physical descriptions, varieties, time and place of growth, more. 2,706 illustrations. xlv + 1,678pp. 8½ × 12¼. 23147-X Cloth. $75.00

DOROTHY AND THE WIZARD IN OZ, L. Frank Baum. Dorothy and the Wizard visit the center of the Earth, where people are vegetables, glass houses grow and Oz characters reappear. Classic sequel to *Wizard of Oz*. 256pp. 5⅜ × 8. 24714-7 Pa. $4.95

SONGS OF EXPERIENCE: Facsimile Reproduction with 26 Plates in Full Color, William Blake. This facsimile of Blake's original "Illuminated Book" reproduces 26 full-color plates from a rare 1826 edition. Includes "The Tyger," "London," "Holy Thursday," and other immortal poems. 26 color plates. Printed text of poems. 48pp. 5¼ × 7. 24636-1 Pa. $3.50

SONGS OF INNOCENCE, William Blake. The first and most popular of Blake's famous "Illuminated Books," in a facsimile edition reproducing all 31 brightly colored plates. Additional printed text of each poem. 64pp. 5¼ × 7. 22764-2 Pa. $3.50

PRECIOUS STONES, Max Bauer. Classic, thorough study of diamonds, rubies, emeralds, garnets, etc.: physical character, occurrence, properties, use, similar topics. 20 plates, 8 in color. 94 figures. 659pp. 6⅛ × 9¼. 21910-0, 21911-9 Pa., Two-vol. set $15.90

ENCYCLOPEDIA OF VICTORIAN NEEDLEWORK, S. F. A. Caulfeild and Blanche Saward. Full, precise descriptions of stitches, techniques for dozens of needlecrafts—most exhaustive reference of its kind. Over 800 figures. Total of 679pp. 8⅛ × 11. Two volumes. Vol. 1 22800-2 Pa. $11.95
Vol. 2 22801-0 Pa. $11.95

THE MARVELOUS LAND OF OZ, L. Frank Baum. Second Oz book, the Scarecrow and Tin Woodman are back with hero named Tip, Oz magic. 136 illustrations. 287pp. 5⅜ × 8½. 20692-0 Pa. $5.95

WILD FOWL DECOYS, Joel Barber. Basic book on the subject, by foremost authority and collector. Reveals history of decoy making and rigging, place in American culture, different kinds of decoys, how to make them, and how to use them. 140 plates. 156pp. 7⅞ × 10¾. 20011-6 Pa. $8.95

HISTORY OF LACE, Mrs. Bury Palliser. Definitive, profusely illustrated chronicle of lace from earliest times to late 19th century. Laces of Italy, Greece, England, France, Belgium, etc. Landmark of needlework scholarship. 266 illustrations. 672pp. 6⅛ × 9¼. 24742-2 Pa. $14.95

ILLUSTRATED GUIDE TO SHAKER FURNITURE, Robert Meader. All furniture and appurtenances, with much on unknown local styles. 235 photos. 146pp. 9 × 12. 22819-3 Pa. $7.95

WHALE SHIPS AND WHALING: A Pictorial Survey, George Francis Dow. Over 200 vintage engravings, drawings, photographs of barks, brigs, cutters, other vessels. Also harpoons, lances, whaling guns, many other artifacts. Comprehensive text by foremost authority. 207 black-and-white illustrations. 288pp. 6 × 9. 24808-9 Pa. $8.95

THE BERTRAMS, Anthony Trollope. Powerful portrayal of blind self-will and thwarted ambition includes one of Trollope's most heartrending love stories. 497pp. 5⅜ × 8½. 25119-5 Pa. $8.95

ADVENTURES WITH A HAND LENS, Richard Headstrom. Clearly written guide to observing and studying flowers and grasses, fish scales, moth and insect wings, egg cases, buds, feathers, seeds, leaf scars, moss, molds, ferns, common crystals, etc.—all with an ordinary, inexpensive magnifying glass. 209 exact line drawings aid in your discoveries. 220pp. 5⅜ × 8½. 23330-8 Pa. $3.95

RODIN ON ART AND ARTISTS, Auguste Rodin. Great sculptor's candid, wide-ranging comments on meaning of art; great artists; relation of sculpture to poetry, painting, music; philosophy of life, more. 76 superb black-and-white illustrations of Rodin's sculpture, drawings and prints. 119pp. 8⅜ × 11¼. 24487-3 Pa. $6.95

FIFTY CLASSIC FRENCH FILMS, 1912–1982: A Pictorial Record, Anthony Slide. Memorable stills from Grand Illusion, Beauty and the Beast, Hiroshima, Mon Amour, many more. Credits, plot synopses, reviews, etc. 160pp. 8¼ × 11. 25256-6 Pa. $11.95

THE PRINCIPLES OF PSYCHOLOGY, William James. Famous long course complete, unabridged. Stream of thought, time perception, memory, experimental methods; great work decades ahead of its time. 94 figures. 1,391pp. 5⅜ × 8½. 20381-6, 20382-4 Pa., Two-vol. set $19.90

BODIES IN A BOOKSHOP, R. T. Campbell. Challenging mystery of blackmail and murder with ingenious plot and superbly drawn characters. In the best tradition of British suspense fiction. 192pp. 5⅜ × 8½. 24720-1 Pa. $3.95

CALLAS: PORTRAIT OF A PRIMA DONNA, George Jellinek. Renowned commentator on the musical scene chronicles incredible career and life of the most controversial, fascinating, influential operatic personality of our time. 64 black-and-white photographs. 416pp. 5⅜ × 8¼. 25047-4 Pa. $7.95

GEOMETRY, RELATIVITY AND THE FOURTH DIMENSION, Rudolph Rucker. Exposition of fourth dimension, concepts of relativity as Flatland characters continue adventures. Popular, easily followed yet accurate, profound. 141 illustrations. 133pp. 5⅜ × 8½. 23400-2 Pa. $3.95

HOUSEHOLD STORIES BY THE BROTHERS GRIMM, with pictures by Walter Crane. 53 classic stories—Rumpelstiltskin, Rapunzel, Hansel and Gretel, the Fisherman and his Wife, Snow White, Tom Thumb, Sleeping Beauty, Cinderella, and so much more—lavishly illustrated with original 19th century drawings. 114 illustrations. x + 269pp. 5⅜ × 8½. 21080-4 Pa. $4.50

SUNDIALS, Albert Waugh. Far and away the best, most thorough coverage of ideas, mathematics concerned, types, construction, adjusting anywhere. Over 100 illustrations. 230pp. $5\frac{3}{8} \times 8\frac{1}{2}$. 22947-5 Pa. $4.50

PICTURE HISTORY OF THE NORMANDIE: With 190 Illustrations, Frank O. Braynard. Full story of legendary French ocean liner: Art Deco interiors, design innovations, furnishings, celebrities, maiden voyage, tragic fire, much more. Extensive text. 144pp. $8\frac{3}{8} \times 11\frac{3}{4}$. 25257-4 Pa. $9.95

THE FIRST AMERICAN COOKBOOK: A Facsimile of "American Cookery," 1796, Amelia Simmons. Facsimile of the first American-written cookbook published in the United States contains authentic recipes for colonial favorites—pumpkin pudding, winter squash pudding, spruce beer, Indian slapjacks, and more. Introductory Essay and Glossary of colonial cooking terms. 80pp. $5\frac{3}{8} \times 8\frac{1}{2}$. 24710-4 Pa. $3.50

101 PUZZLES IN THOUGHT AND LOGIC, C. R. Wylie, Jr. Solve murders and robberies, find out which fishermen are liars, how a blind man could possibly identify a color—purely by your own reasoning! 107pp. $5\frac{3}{8} \times 8\frac{1}{2}$. 20367-0 Pa. $2.50

THE BOOK OF WORLD-FAMOUS MUSIC—CLASSICAL, POPULAR AND FOLK, James J. Fuld. Revised and enlarged republication of landmark work in musico-bibliography. Full information about nearly 1,000 songs and compositions including first lines of music and lyrics. New supplement. Index. 800pp. $5\frac{3}{8} \times 8\frac{1}{4}$. 24857-7 Pa. $14.95

ANTHROPOLOGY AND MODERN LIFE, Franz Boas. Great anthropologist's classic treatise on race and culture. Introduction by Ruth Bunzel. Only inexpensive paperback edition. 255pp. $5\frac{3}{8} \times 8\frac{1}{2}$. 25245-0 Pa. $5.95

THE TALE OF PETER RABBIT, Beatrix Potter. The inimitable Peter's terrifying adventure in Mr. McGregor's garden, with all 27 wonderful, full-color Potter illustrations. 55pp. $4\frac{1}{4} \times 5\frac{1}{2}$. (Available in U.S. only) 22827-4 Pa. $1.75

THREE PROPHETIC SCIENCE FICTION NOVELS, H. G. Wells. *When the Sleeper Wakes, A Story of the Days to Come* and *The Time Machine* (full version). 335pp. $5\frac{3}{8} \times 8\frac{1}{2}$. (Available in U.S. only) 20605-X Pa. $5.95

APICIUS COOKERY AND DINING IN IMPERIAL ROME, edited and translated by Joseph Dommers Vehling. Oldest known cookbook in existence offers readers a clear picture of what foods Romans ate, how they prepared them, etc. 49 illustrations. 301pp. $6\frac{1}{8} \times 9\frac{1}{4}$. 23563-7 Pa. $6.50

SHAKESPEARE LEXICON AND QUOTATION DICTIONARY, Alexander Schmidt. Full definitions, locations, shades of meaning of every word in plays and poems. More than 50,000 exact quotations. 1,485pp. $6\frac{1}{2} \times 9\frac{1}{4}$.
22726-X, 22727-8 Pa., Two-vol. set $27.90

THE WORLD'S GREAT SPEECHES, edited by Lewis Copeland and Lawrence W. Lamm. Vast collection of 278 speeches from Greeks to 1970. Powerful and effective models; unique look at history. 842pp. $5\frac{3}{8} \times 8\frac{1}{2}$. 20468-5 Pa. $11.95

THE BLUE FAIRY BOOK, Andrew Lang. The first, most famous collection, with many familiar tales: Little Red Riding Hood, Aladdin and the Wonderful Lamp, Puss in Boots, Sleeping Beauty, Hansel and Gretel, Rumpelstiltskin; 37 in all. 138 illustrations. 390pp. 5⅜ × 8½. 21437-0 Pa. $5.95

THE STORY OF THE CHAMPIONS OF THE ROUND TABLE, Howard Pyle. Sir Launcelot, Sir Tristram and Sir Percival in spirited adventures of love and triumph retold in Pyle's inimitable style. 50 drawings, 31 full-page. xviii + 329pp. 6½ × 9¼. 21883-X Pa. $6.95

AUDUBON AND HIS JOURNALS, Maria Audubon. Unmatched two-volume portrait of the great artist, naturalist and author contains his journals, an excellent biography by his granddaughter, expert annotations by the noted ornithologist, Dr. Elliott Coues, and 37 superb illustrations. Total of 1,200pp. 5⅜ × 8.
Vol. I 25143-8 Pa. $8.95
Vol. II 25144-6 Pa. $8.95

GREAT DINOSAUR HUNTERS AND THEIR DISCOVERIES, Edwin H. Colbert. Fascinating, lavishly illustrated chronicle of dinosaur research, 1820's to 1960. Achievements of Cope, Marsh, Brown, Buckland, Mantell, Huxley, many others. 384pp. 5¼ × 8¼. 24701-5 Pa. $6.95

THE TASTEMAKERS, Russell Lynes. Informal, illustrated social history of American taste 1850's–1910's. First popularized categories Highbrow, Lowbrow, Middlebrow. 129 illustrations. New (1979) afterword. 384pp. 6 × 9.
23993-4 Pa. $6.95

DOUBLE CROSS PURPOSES, Ronald A. Knox. A treasure hunt in the Scottish Highlands, an old map, unidentified corpse, surprise discoveries keep reader guessing in this cleverly intricate tale of financial skullduggery. 2 black-and-white maps. 320pp. 5⅜ × 8½. (Available in U.S. only) 25032-6 Pa. $5.95

AUTHENTIC VICTORIAN DECORATION AND ORNAMENTATION IN FULL COLOR: 46 Plates from "Studies in Design," Christopher Dresser. Superb full-color lithographs reproduced from rare original portfolio of a major Victorian designer. 48pp. 9¼ × 12¼. 25083-0 Pa. $7.95

PRIMITIVE ART, Franz Boas. Remains the best text ever prepared on subject, thoroughly discussing Indian, African, Asian, Australian, and, especially, Northern American primitive art. Over 950 illustrations show ceramics, masks, totem poles, weapons, textiles, paintings, much more. 376pp. 5⅜ × 8. 20025-6 Pa. $6.95

SIDELIGHTS ON RELATIVITY, Albert Einstein. Unabridged republication of two lectures delivered by the great physicist in 1920-21. *Ether and Relativity* and *Geometry and Experience*. Elegant ideas in non-mathematical form, accessible to intelligent layman. vi + 56pp. 5⅜ × 8½. 24511-X Pa. $2.95

THE WIT AND HUMOR OF OSCAR WILDE, edited by Alvin Redman. More than 1,000 ripostes, paradoxes, wisecracks: Work is the curse of the drinking classes, I can resist everything except temptation, etc. 258pp. 5⅜ × 8½. 20602-5 Pa. $4.50

ADVENTURES WITH A MICROSCOPE, Richard Headstrom. 59 adventures with clothing fibers, protozoa, ferns and lichens, roots and leaves, much more. 142 illustrations. 232pp. 5⅜ × 8½. 23471-1 Pa. $3.95

CATALOG OF DOVER BOOKS

PLANTS OF THE BIBLE, Harold N. Moldenke and Alma L. Moldenke. Standard reference to all 230 plants mentioned in Scriptures. Latin name, biblical reference, uses, modern identity, much more. Unsurpassed encyclopedic resource for scholars, botanists, nature lovers, students of Bible. Bibliography. Indexes. 123 black-and-white illustrations. 384pp. 6 × 9.   25069-5 Pa. $8.95

FAMOUS AMERICAN WOMEN: A Biographical Dictionary from Colonial Times to the Present, Robert McHenry, ed. From Pocahontas to Rosa Parks, 1,035 distinguished American women documented in separate biographical entries. Accurate, up-to-date data, numerous categories, spans 400 years. Indices. 493pp. 6½ × 9¼.   24523-3 Pa. $9.95

THE FABULOUS INTERIORS OF THE GREAT OCEAN LINERS IN HISTORIC PHOTOGRAPHS, William H. Miller, Jr. Some 200 superb photographs capture exquisite interiors of world's great "floating palaces"—1890's to 1980's: Titanic, Ile de France, Queen Elizabeth, United States, Europa, more. Approx. 200 black-and-white photographs. Captions. Text. Introduction. 160pp. 8⅜ × 11¼.
                                                                        24756-2 Pa. $9.95

THE GREAT LUXURY LINERS, 1927–1954: A Photographic Record, William H. Miller, Jr. Nostalgic tribute to heyday of ocean liners. 186 photos of Ile de France, Normandie, Leviathan, Queen Elizabeth, United States, many others. Interior and exterior views. Introduction. Captions. 160pp. 9 × 12.
                                                                        24056-8 Pa. $9.95

A NATURAL HISTORY OF THE DUCKS, John Charles Phillips. Great landmark of ornithology offers complete detailed coverage of nearly 200 species and subspecies of ducks: gadwall, sheldrake, merganser, pintail, many more. 74 full-color plates, 102 black-and-white. Bibliography. Total of 1,920pp. 8⅜ × 11¼.
                                                25141-1, 25142-X Cloth. Two-vol. set $100.00

THE SEAWEED HANDBOOK: An Illustrated Guide to Seaweeds from North Carolina to Canada, Thomas F. Lee. Concise reference covers 78 species. Scientific and common names, habitat, distribution, more. Finding keys for easy identification. 224pp. 5⅜ × 8½.   25215-9 Pa. $5.95

THE TEN BOOKS OF ARCHITECTURE: The 1755 Leoni Edition, Leon Battista Alberti. Rare classic helped introduce the glories of ancient architecture to the Renaissance. 68 black-and-white plates. 336pp. 8⅜ × 11¼.   25239-6 Pa. $14.95

MISS MACKENZIE, Anthony Trollope. Minor masterpieces by Victorian master unmasks many truths about life in 19th-century England. First inexpensive edition in years. 392pp. 5⅜ × 8½.   25201-9 Pa. $7.95

THE RIME OF THE ANCIENT MARINER, Gustave Doré, Samuel Taylor Coleridge. Dramatic engravings considered by many to be his greatest work. The terrifying space of the open sea, the storms and whirlpools of an unknown ocean, the ice of Antarctica, more—all rendered in a powerful, chilling manner. Full text. 38 plates. 77pp. 9¼ × 12.   22305-1 Pa. $4.95

THE EXPEDITIONS OF ZEBULON MONTGOMERY PIKE, Zebulon Montgomery Pike. Fascinating first-hand accounts (1805-6) of exploration of Mississippi River, Indian wars, capture by Spanish dragoons, much more. 1,088pp. 5⅜ × 8½.   25254-X, 25255-8 Pa. Two-vol. set $23.90

A CONCISE HISTORY OF PHOTOGRAPHY: Third Revised Edition, Helmut Gernsheim. Best one-volume history—camera obscura, photochemistry, daguerreotypes, evolution of cameras, film, more. Also artistic aspects—landscape, portraits, fine art, etc. 281 black-and-white photographs. 26 in color. 176pp. 8⅜ × 11¼. 25128-4 Pa. $12.95

THE DORÉ BIBLE ILLUSTRATIONS, Gustave Doré. 241 detailed plates from the Bible: the Creation scenes, Adam and Eve, Flood, Babylon, battle sequences, life of Jesus, etc. Each plate is accompanied by the verses from the King James version of the Bible. 241pp. 9 × 12. 23004-X Pa. $8.95

HUGGER-MUGGER IN THE LOUVRE, Elliot Paul. Second Homer Evans mystery-comedy. Theft at the Louvre involves sleuth in hilarious, madcap caper. "A knockout."—Books. 336pp. 5⅜ × 8½. 25185-3 Pa. $5.95

FLATLAND, E. A. Abbott. Intriguing and enormously popular science-fiction classic explores the complexities of trying to survive as a two-dimensional being in a three-dimensional world. Amusingly illustrated by the author. 16 illustrations. 103pp. 5⅜ × 8½. 20001-9 Pa. $2.25

THE HISTORY OF THE LEWIS AND CLARK EXPEDITION, Meriwether Lewis and William Clark, edited by Elliott Coues. Classic edition of Lewis and Clark's day-by-day journals that later became the basis for U.S. claims to Oregon and the West. Accurate and invaluable geographical, botanical, biological, meteorological and anthropological material. Total of 1,508pp. 5⅜ × 8½. 21268-8, 21269-6, 21270-X Pa. Three-vol. set $25.50

LANGUAGE, TRUTH AND LOGIC, Alfred J. Ayer. Famous, clear introduction to Vienna, Cambridge schools of Logical Positivism. Role of philosophy, elimination of metaphysics, nature of analysis, etc. 160pp. 5⅜ × 8½. (Available in U.S. and Canada only) 20010-8 Pa. $2.95

MATHEMATICS FOR THE NONMATHEMATICIAN, Morris Kline. Detailed, college-level treatment of mathematics in cultural and historical context, with numerous exercises. For liberal arts students. Preface. Recommended Reading Lists. Tables. Index. Numerous black-and-white figures. xvi + 641pp. 5⅜ × 8½. 24823-2 Pa. $11.95

28 SCIENCE FICTION STORIES, H. G. Wells. Novels, *Star Begotten* and *Men Like Gods*, plus 26 short stories: "Empire of the Ants," "A Story of the Stone Age," "The Stolen Bacillus," "In the Abyss," etc. 915pp. 5⅜ × 8½. (Available in U.S. only) 20265-8 Cloth. $10.95

HANDBOOK OF PICTORIAL SYMBOLS, Rudolph Modley. 3,250 signs and symbols, many systems in full; official or heavy commercial use. Arranged by subject. Most in Pictorial Archive series. 143pp. 8⅜ × 11. 23357-X Pa. $5.95

INCIDENTS OF TRAVEL IN YUCATAN, John L. Stephens. Classic (1843) exploration of jungles of Yucatan, looking for evidences of Maya civilization. Travel adventures, Mexican and Indian culture, etc. Total of 669pp. 5⅜ × 8½. 20926-1, 20927-X Pa., Two-vol. set $9.90

DEGAS: An Intimate Portrait, Ambroise Vollard. Charming, anecdotal memoir by famous art dealer of one of the greatest 19th-century French painters. 14 black-and-white illustrations. Introduction by Harold L. Van Doren. 96pp. 5⅜ × 8½.
25131-4 Pa. $3.95

PERSONAL NARRATIVE OF A PILGRIMAGE TO ALMANDINAH AND MECCAH, Richard Burton. Great travel classic by remarkably colorful personality. Burton, disguised as a Moroccan, visited sacred shrines of Islam, narrowly escaping death. 47 illustrations. 959pp. 5⅜ × 8½. 21217-3, 21218-1 Pa., Two-vol. set $19.90

PHRASE AND WORD ORIGINS, A. H. Holt. Entertaining, reliable, modern study of more than 1,200 colorful words, phrases, origins and histories. Much unexpected information. 254pp. 5⅜ × 8½. 20758-7 Pa. $4.95

THE RED THUMB MARK, R. Austin Freeman. In this first Dr. Thorndyke case, the great scientific detective draws fascinating conclusions from the nature of a single fingerprint. Exciting story, authentic science. 320pp. 5⅜ × 8½. (Available in U.S. only) 25210-8 Pa. $5.95

AN EGYPTIAN HIEROGLYPHIC DICTIONARY, E. A. Wallis Budge. Monumental work containing about 25,000 words or terms that occur in texts ranging from 3000 B.C. to 600 A.D. Each entry consists of a transliteration of the word, the word in hieroglyphs, and the meaning in English. 1,314pp. 6⅜ × 10.
23615-3, 23616-1 Pa., Two-vol. set $27.90

THE COMPLEAT STRATEGYST: Being a Primer on the Theory of Games of Strategy, J. D. Williams. Highly entertaining classic describes, with many illustrated examples, how to select best strategies in conflict situations. Prefaces. Appendices. xvi + 268pp. 5⅜ × 8½. 25101-2 Pa. $5.95

THE ROAD TO OZ, L. Frank Baum. Dorothy meets the Shaggy Man, little Button-Bright and the Rainbow's beautiful daughter in this delightful trip to the magical Land of Oz. 272pp. 5⅜ × 8. 25208-6 Pa. $4.95

POINT AND LINE TO PLANE, Wassily Kandinsky. Seminal exposition of role of point, line, other elements in non-objective painting. Essential to understanding 20th-century art. 127 illustrations. 192pp. 6½ × 9¼. 23808-3 Pa. $4.50

LADY ANNA, Anthony Trollope. Moving chronicle of Countess Lovel's bitter struggle to win for herself and daughter Anna their rightful rank and fortune—perhaps at cost of sanity itself. 384pp. 5⅜ × 8½. 24669-8 Pa. $6.95

EGYPTIAN MAGIC, E. A. Wallis Budge. Sums up all that is known about magic in Ancient Egypt: the role of magic in controlling the gods, powerful amulets that warded off evil spirits, scarabs of immortality, use of wax images, formulas and spells, the secret name, much more. 253pp. 5⅜ × 8½. 22681-6 Pa. $4.00

THE DANCE OF SIVA, Ananda Coomaraswamy. Preeminent authority unfolds the vast metaphysic of India: the revelation of her art, conception of the universe, social organization, etc. 27 reproductions of art masterpieces. 192pp. 5⅜ × 8½.
24817-8 Pa. $5.95

CHRISTMAS CUSTOMS AND TRADITIONS, Clement A. Miles. Origin, evolution, significance of religious, secular practices. Caroling, gifts, yule logs, much more. Full, scholarly yet fascinating; non-sectarian. 400pp. 5⅜ × 8½.
23354-5 Pa. $6.50

THE HUMAN FIGURE IN MOTION, Eadweard Muybridge. More than 4,500 stopped-action photos, in action series, showing undraped men, women, children jumping, lying down, throwing, sitting, wrestling, carrying, etc. 390pp. 7⅞ × 10⅝.
20204-6 Cloth. $21.95

THE MAN WHO WAS THURSDAY, Gilbert Keith Chesterton. Witty, fast-paced novel about a club of anarchists in turn-of-the-century London. Brilliant social, religious, philosophical speculations. 128pp. 5⅜ × 8½.
25121-7 Pa. $3.95

A CEZANNE SKETCHBOOK: Figures, Portraits, Landscapes and Still Lifes, Paul Cezanne. Great artist experiments with tonal effects, light, mass, other qualities in over 100 drawings. A revealing view of developing master painter, precursor of Cubism. 102 black-and-white illustrations. 144pp. 8¾ × 6⅜.
24790-2 Pa. $5.95

AN ENCYCLOPEDIA OF BATTLES: Accounts of Over 1,560 Battles from 1479 B.C. to the Present, David Eggenberger. Presents essential details of every major battle in recorded history, from the first battle of Megiddo in 1479 B.C. to Grenada in 1984. List of Battle Maps. New Appendix covering the years 1967–1984. Index. 99 illustrations. 544pp. 6½ × 9¼.
24913-1 Pa. $14.95

AN ETYMOLOGICAL DICTIONARY OF MODERN ENGLISH, Ernest Weekley. Richest, fullest work, by foremost British lexicographer. Detailed word histories. Inexhaustible. Total of 856pp. 6½ × 9¼.
21873-2, 21874-0 Pa., Two-vol. set $17.00

WEBSTER'S AMERICAN MILITARY BIOGRAPHIES, edited by Robert McHenry. Over 1,000 figures who shaped 3 centuries of American military history. Detailed biographies of Nathan Hale, Douglas MacArthur, Mary Hallaren, others. Chronologies of engagements, more. Introduction. Addenda. 1,033 entries in alphabetical order. xi + 548pp. 6½ × 9¼. (Available in U.S. only)
24758-9 Pa. $11.95

LIFE IN ANCIENT EGYPT, Adolf Erman. Detailed older account, with much not in more recent books: domestic life, religion, magic, medicine, commerce, and whatever else needed for complete picture. Many illustrations. 597pp. 5⅜ × 8½.
22632-8 Pa. $8.50

HISTORIC COSTUME IN PICTURES, Braun & Schneider. Over 1,450 costumed figures shown, covering a wide variety of peoples: kings, emperors, nobles, priests, servants, soldiers, scholars, townsfolk, peasants, merchants, courtiers, cavaliers, and more. 256pp. 8⅜ × 11¼.
23150-X Pa. $7.95

THE NOTEBOOKS OF LEONARDO DA VINCI, edited by J. P. Richter. Extracts from manuscripts reveal great genius; on painting, sculpture, anatomy, sciences, geography, etc. Both Italian and English. 186 ms. pages reproduced, plus 500 additional drawings, including studies for *Last Supper, Sforza* monument, etc. 860pp. 7⅞ × 10¾. (Available in U.S. only) 22572-0, 22573-9 Pa., Two-vol. set $25.90

THE ART NOUVEAU STYLE BOOK OF ALPHONSE MUCHA: All 72 Plates from "Documents Decoratifs" in Original Color, Alphonse Mucha. Rare copyright-free design portfolio by high priest of Art Nouveau. Jewelry, wallpaper, stained glass, furniture, figure studies, plant and animal motifs, etc. Only complete one-volume edition. 80pp. 9⅜ × 12¼. 24044-4 Pa. $8.95

ANIMALS: 1,419 COPYRIGHT-FREE ILLUSTRATIONS OF MAMMALS, BIRDS, FISH, INSECTS, ETC., edited by Jim Harter. Clear wood engravings present, in extremely lifelike poses, over 1,000 species of animals. One of the most extensive pictorial sourcebooks of its kind. Captions. Index. 284pp. 9 × 12. 23766-4 Pa. $9.95

OBELISTS FLY HIGH, C. Daly King. Masterpiece of American detective fiction, long out of print, involves murder on a 1935 transcontinental flight—"a very thrilling story"—NY Times. Unabridged and unaltered republication of the edition published by William Collins Sons & Co. Ltd., London, 1935. 288pp. 5⅜ × 8½. (Available in U.S. only) 25036-9 Pa. $4.95

VICTORIAN AND EDWARDIAN FASHION: A Photographic Survey, Alison Gernsheim. First fashion history completely illustrated by contemporary photographs. Full text plus 235 photos, 1840–1914, in which many celebrities appear. 240pp. 6½ × 9¼. 24205-6 Pa. $6.00

THE ART OF THE FRENCH ILLUSTRATED BOOK, 1700–1914, Gordon N. Ray. Over 630 superb book illustrations by Fragonard, Delacroix, Daumier, Doré, Grandville, Manet, Mucha, Steinlen, Toulouse-Lautrec and many others. Preface. Introduction. 633 halftones. Indices of artists, authors & titles, binders and provenances. Appendices. Bibliography. 608pp. 8⅜ × 11¼. 25086-5 Pa. $24.95

THE WONDERFUL WIZARD OF OZ, L. Frank Baum. Facsimile in full color of America's finest children's classic. 143 illustrations by W. W. Denslow. 267pp. 5⅜ × 8½. 20691-2 Pa. $5.95

FRONTIERS OF MODERN PHYSICS: New Perspectives on Cosmology, Relativity, Black Holes and Extraterrestrial Intelligence, Tony Rothman, et al. For the intelligent layman. Subjects include: cosmological models of the universe; black holes; the neutrino; the search for extraterrestrial intelligence. Introduction. 46 black-and-white illustrations. 192pp. 5⅜ × 8½. 24587-X Pa. $6.95

THE FRIENDLY STARS, Martha Evans Martin & Donald Howard Menzel. Classic text marshalls the stars together in an engaging, non-technical survey, presenting them as sources of beauty in night sky. 23 illustrations. Foreword. 2 star charts. Index. 147pp. 5⅜ × 8½. 21099-5 Pa. $3.50

FADS AND FALLACIES IN THE NAME OF SCIENCE, Martin Gardner. Fair, witty appraisal of cranks, quacks, and quackeries of science and pseudoscience: hollow earth, Velikovsky, orgone energy, Dianetics, flying saucers, Bridey Murphy, food and medical fads, etc. Revised, expanded In the Name of Science. "A very able and even-tempered presentation."—The New Yorker. 363pp. 5⅜ × 8. 20394-8 Pa. $6.50

ANCIENT EGYPT: ITS CULTURE AND HISTORY, J. E Manchip White. From pre-dynastics through Ptolemies: society, history, political structure, religion, daily life, literature, cultural heritage. 48 plates. 217pp. 5⅜ × 8½. 22548-8 Pa. $4.95

CATALOG OF DOVER BOOKS

SIR HARRY HOTSPUR OF HUMBLETHWAITE, Anthony Trollope. Incisive, unconventional psychological study of a conflict between a wealthy baronet, his idealistic daughter, and their scapegrace cousin. The 1870 novel in its first inexpensive edition in years. 250pp. 5⅜ × 8½.   24953-0 Pa. $5.95

LASERS AND HOLOGRAPHY, Winston E. Kock. Sound introduction to burgeoning field, expanded (1981) for second edition. Wave patterns, coherence, lasers, diffraction, zone plates, properties of holograms, recent advances. 84 illustrations. 160pp. 5⅜ × 8¼. (Except in United Kingdom)   24041-X Pa. $3.50

INTRODUCTION TO ARTIFICIAL INTELLIGENCE: SECOND, EN-LARGED EDITION, Philip C. Jackson, Jr. Comprehensive survey of artificial intelligence—the study of how machines (computers) can be made to act intelligently. Includes introductory and advanced material. Extensive notes updating the main text. 132 black-and-white illustrations. 512pp. 5⅜ × 8½.   24864-X Pa. $8.95

HISTORY OF INDIAN AND INDONESIAN ART, Ananda K. Coomaraswamy. Over 400 illustrations illuminate classic study of Indian art from earliest Harappa finds to early 20th century. Provides philosophical, religious and social insights. 304pp. 6⅜ × 9⅜.   25005-9 Pa. $8.95

THE GOLEM, Gustav Meyrink. Most famous supernatural novel in modern European literature, set in Ghetto of Old Prague around 1890. Compelling story of mystical experiences, strange transformations, profound terror. 13 black-and-white illustrations. 224pp. 5⅜ × 8½. (Available in U.S. only)   25025-3 Pa. $5.95

ARMADALE, Wilkie Collins. Third great mystery novel by the author of *The Woman in White* and *The Moonstone*. Original magazine version with 40 illustrations. 597pp. 5⅜ × 8½.   23429-0 Pa. $9.95

PICTORIAL ENCYCLOPEDIA OF HISTORIC ARCHITECTURAL PLANS, DETAILS AND ELEMENTS: With 1,880 Line Drawings of Arches, Domes, Doorways, Facades, Gables, Windows, etc., John Theodore Haneman. Sourcebook of inspiration for architects, designers, others. Bibliography. Captions. 141pp. 9 × 12.   24605-1 Pa. $6.95

BENCHLEY LOST AND FOUND, Robert Benchley. Finest humor from early 30's, about pet peeves, child psychologists, post office and others. Mostly unavailable elsewhere. 73 illustrations by Peter Arno and others. 183pp. 5⅜ × 8½.   22410-4 Pa. $3.95

ERTÉ GRAPHICS, Erté. Collection of striking color graphics: *Seasons, Alphabet, Numerals, Aces* and *Precious Stones*. 50 plates, including 4 on covers. 48pp. 9⅜ × 12¼.   23580-7 Pa. $6.95

THE JOURNAL OF HENRY D. THOREAU, edited by Bradford Torrey, F. H. Allen. Complete reprinting of 14 volumes, 1837-61, over two million words; the sourcebooks for *Walden*, etc. Definitive. All original sketches, plus 75 photographs. 1,804pp. 8½ × 12¼.   20312-3, 20313-1 Cloth., Two-vol. set $80.00

CASTLES: THEIR CONSTRUCTION AND HISTORY, Sidney Toy. Traces castle development from ancient roots. Nearly 200 photographs and drawings illustrate moats, keeps, baileys, many other features. Caernarvon, Dover Castles, Hadrian's Wall, Tower of London, dozens more. 256pp. 5⅜ × 8¼.   24898-4 Pa. $5.95

CATALOG OF DOVER BOOKS

AMERICAN CLIPPER SHIPS: 1833–1858, Octavius T. Howe & Frederick C. Matthews. Fully-illustrated, encyclopedic review of 352 clipper ships from the period of America's greatest maritime supremacy. Introduction. 109 halftones. 5 black-and-white line illustrations. Index. Total of 928pp. 5⅜ × 8½.
25115-2, 25116-0 Pa., Two-vol. set $17.90

TOWARDS A NEW ARCHITECTURE, Le Corbusier. Pioneering manifesto by great architect, near legendary founder of "International School." Technical and aesthetic theories, views on industry, economics, relation of form to function, "mass-production spirit," much more. Profusely illustrated. Unabridged translation of 13th French edition. Introduction by Frederick Etchells. 320pp. 6⅛ × 9¼. (Available in U.S. only)
25023-7 Pa. $8.95

THE BOOK OF KELLS, edited by Blanche Cirker. Inexpensive collection of 32 full-color, full-page plates from the greatest illuminated manuscript of the Middle Ages, painstakingly reproduced from rare facsimile edition. Publisher's Note. Captions. 32pp. 9⅜ × 12¼.
24345-1 Pa. $4.95

BEST SCIENCE FICTION STORIES OF H. G. WELLS, H. G. Wells. Full novel *The Invisible Man*, plus 17 short stories: "The Crystal Egg," "Aepyornis Island," "The Strange Orchid," etc. 303pp. 5⅜ × 8½. (Available in U.S. only)
21531-8 Pa. $4.95

AMERICAN SAILING SHIPS: Their Plans and History, Charles G. Davis. Photos, construction details of schooners, frigates, clippers, other sailcraft of 18th to early 20th centuries—plus entertaining discourse on design, rigging, nautical lore, much more. 137 black-and-white illustrations. 240pp. 6⅛ × 9¼.
24658-2 Pa. $5.95

ENTERTAINING MATHEMATICAL PUZZLES, Martin Gardner. Selection of author's favorite conundrums involving arithmetic, money, speed, etc., with lively commentary. Complete solutions. 112pp. 5⅜ × 8½. 25211-6 Pa. $2.95

THE WILL TO BELIEVE, HUMAN IMMORTALITY, William James. Two books bound together. Effect of irrational on logical, and arguments for human immortality. 402pp. 5⅜ × 8½. 20291-7 Pa. $7.50

THE HAUNTED MONASTERY and THE CHINESE MAZE MURDERS, Robert Van Gulik. 2 full novels by Van Gulik continue adventures of Judge Dee and his companions. An evil Taoist monastery, seemingly supernatural events; overgrown topiary maze that hides strange crimes. Set in 7th-century China. 27 illustrations. 328pp. 5⅜ × 8½. 23502-5 Pa. $5.95

CELEBRATED CASES OF JUDGE DEE (DEE GOONG AN), translated by Robert Van Gulik. Authentic 18th-century Chinese detective novel; Dee and associates solve three interlocked cases. Led to Van Gulik's own stories with same characters. Extensive introduction. 9 illustrations. 237pp. 5⅜ × 8½.
23337-5 Pa. $4.95

Prices subject to change without notice.
Available at your book dealer or write for free catalog to Dept. GI, Dover Publications, Inc., 31 East 2nd St., Mineola, N.Y. 11501. Dover publishes more than 175 books each year on science, elementary and advanced mathematics, biology, music, art, literary history, social sciences and other areas.